Multiscale Modeling and Simulation of Shock Wave-Induced Failure in Materials Science

Martin Oliver Steinhauser

Multiscale Modeling and Simulation of Shock Wave-Induced Failure in Materials Science

 Springer Spektrum

Martin Oliver Steinhauser
Freiburg im Breisgau, Germany

Habilitation treatise University of Basel, Faculty of Science, Department of Chemistry,
Basel, Switzerland, 2017

ISBN 978-3-658-21133-2   ISBN 978-3-658-21134-9   (eBook)
https://doi.org/10.1007/978-3-658-21134-9

Library of Congress Control Number: 2018934376

Springer Spektrum

Printed on acid-free paper

This Springer Spektrum imprint is published by Springer Nature
The registered company is Springer Fachmedien Wiesbaden GmbH
The registered company address is: Abraham-Lincoln-Str. 46, 65189 Wiesbaden, Germany

To my family
Katrin,
Pia,
Sven,
Charly,
and Micky

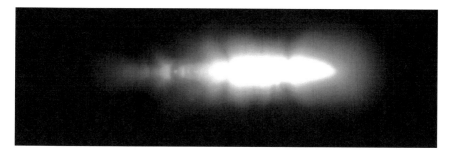

Shock wave, generated by a spark channel during optical breakdown
of a laser in air. High speed photography image.
© *Martin O. Steinhauser.*

Preface

The rapid progress in the development of fast, parallelized computing systems has made it possible to routinely model and simulate physical processes on computers. One major advantage of computer simulations is that they allow for studying physical processes that are very hard or impossible to be studied in real laboratory experiments. In fact, nowadays, computer simulations complement and sometimes even substitute experiments, thus enhancing and also accelerating our understanding of material properties on all relevant length- and time scales.

This monograph is a revised, slightly enhanced and updated version of my Habilitation treatise which I submitted successfully at the Faculty of Science of the University of Basel in Switzerland in the year 2016. It covers a basic introduction to the area of shock wave physics and computational multiscale modeling in materials science. Several state-of-the-art research examples are provided from hard matter and soft matter physics including the description of several major achievements such as a new way of coupling the atomistic with the continuum domain and the successful destruction of tumor cells based on laser-induced shock wave treatment of U87 glioblastoma human brain tumor cells. I hope, this book will be of use for all scientists who want to enter the field of computational multiscale materials science with a particular interest in material behavior at very high strain rates governed by shock wave physics. I think, the book will provide several ideas and stimuli not only for advanced graduate or PhD students but also for the practitioner in that it gives a concise summary and overview of the field along with state-of-the-art applications. The book also contains more than 250 topical references which will help the reader to further advance in these research areas.

I would like to thank Carina Berg and Anita Wilke of Springer Verlag for the swift and easy collaboration in all aspects of editing and book production.

A lot of thanks go to many of my former students, postdocs and co-workers over the years at the Fraunhofer Ernst-Mach-Institut, EMI in Freiburg, Germany. I would like to point out and thank Kai Grass, Julian Schneider and Tobias Pardowitz who worked with me on coarse-graining and computational multiscale modeling of soft matter systems discussed in Chaps. 3 and 5. I would also like to thank Tanja Schindler who worked with me as her advisor of her diploma thesis in mathematics on the topic of numerical modeling of shock wave effects in biomembranes and cells which are presented in Chaps. 5 and 6. Dr. Paul Liedekerke who joined my former Fraunhofer research group *Shock waves in Soft Biological Matter* as a postdoc for one year, assisted me with several modeling approaches for cellular structures. PD Dr. Jens Osterholz helped me with PDV measurements in my laser lab which are discussed shortly in Chap. 6. With Dr. Martin Kühn I learned the mathematical theory of power diagrams, developed numerical methods for the simulation of shock waves in granular materials and conducted impact experiments with high-performance ceramics which are described in Chap. 4.

My special thanks go to Dr. Mischa Schmidt who joined my research group as a postdoctoral researcher conducting cancer cell experiments with me both in the cell lab of the University Hospital in Freiburg and in my laser lab that I had established from scratch at EMI. Some of the results of this research are described in Chap. 6. I would also like to thank Prof. Dr. med. Guido Nikkhah, former director of the Department of Neurosurgery at the University Hospital in Freiburg, for collaborating with me, for very helpful discussions concerning shock wave effects in cell cultures, and for letting me use his cell culture lab facilities. I also had very useful and interesting discussions with Prof. Dr. med. Jaroslaw Maciaczyk and Dr. Ulf Kahlert of the brain tumor stem cell research group in Freiburg.

At several research visits in labs around the world and also at many international conferences I benefited from the knowledge of other researchers while having interesting scholarly debates. Among the many individuals I would like to point out are Prof. Constantin-C. Coussios (Director of the Institute of Biomedical Engineering at Oxford University, UK), and Prof. Roberto Contro (Head of the Laboratory of Biological Structure Mechanics at the Politecnico di Milano, Italy). I also profited from discussions on shock waves in general and their effects in tissue and cells in particular with Prof. Josef Käs (Head of the Soft Matter Physics Division of the Universität Leipzig) and with Prof. Liangchi Zhang who invited me to be guest scientist at his Laboratory for Precision and Nano Processing at the University of New South Wales in Sydney, Australia. I am also thankful to Prof. Klaus Kroy of the Soft Condensed Matter Theory Group at the Universität Leipzig, who invited me as guest speaker for the Graduate School "Building with Molecules and Nano-objects" (BuildMoNa). This was a great opportunity for me to engage in fruitful discussions on theoretical issues about scaling laws in biological, cellular systems with other lecturers and aspiring PhD students.

Special thanks go to Prof. Thomas Lengauer from the Max-Planck-Institut für Informatik who invited me on behalf of the *Leopoldina* – the German National Academy of Sciences – and the Brasilian Academy of Sciences, to be part of the German delegation in a one week scientific workshop on multiscale methods in Rio de Janeiro and São Paulo, where I had the opportunity to present a keynote and exchange ideas on multiscale materials research in soft and hard matter.

At the University of Basel, I'd like to thank my colleagues Prof. Markus Meuwly and Prof. Stefan Willitsch who made me feel welcome in the Department of Chemistry, where I have been teaching many Quantum Chemistry and Computational Science courses on all levels (bachelor, graduate and PhD) during the last 6 years. Teaching these courses besides my actual work at a Fraunhofer research institute, where most of the working hours cannot be devoted to pure and free research, has been very demanding at times, but has also been (and still is) a source of great pleasure for me.

Above all, I thank my wife Katrin and my children Pia and Sven for their patience and understanding, and for supporting me when I was working day and night in my leisure time, preparing this work which is now finally turned into a full-blown monograph publication.

I'd appreciate any positive, productive comments or substantive discussions. I can be reached via eMail at martin.steinhauser@unibas.ch, martin.steinhauser@emi.fraunhofer.de or via Research Gate at the address http://www.researchgate.net.

Martin Oliver Steinhauser
Freiburg im Breisgau, Germany
and Basel, Switzerland
January 2018

Contents

List of Figures

List of Tables

Part I.

Shock wave physics, multiscale modeling and simulation

1. Introduction

This work deals with several aspects of multiscale materials modeling and simulation in applied materials research and fundamental science. The applications for multiscale modeling presented here are shock wave phenomena in condensed matter (with the example application of ceramics) and in soft matter systems (with the examples of polymers and phospholipid macromolecules). We present research results based on multiscale modeling and computer simulations, as well as shock wave experiments with hard matter (ceramics) and soft matter (U87 glioblastoma tumor cells). For the latter, we tested successfully the hypothesis that mechanical shock waves can have a destructive biological effect on this type of human brain tumor cells.

In essence, I demonstrate in this work that the physics of shock waves has grown to become an important interdisciplinary field of research with relevance to applied materials science and to fundamental science. At the same time progress in computer hard- and software development have boosted new ideas in the field of multiscale modeling and simulation. Bridging the scales in a theoretical-numerical description of materials has become a prime challenge in science and technology.

The importance of multiscale modeling was finally recognized by a 2013 Nobel Prize in Chemistry awarded to Martin Karplus, Michael Levitt and Arieh Warshel "for the development of multiscale models for complex chemical systems." One key challenge for multiscale modeling is scale bridging, for example, passing information about behavior of atoms described at the quantum scale to the next higher atomic scale. It might be tempting to contemplate simulations only at the most fundamental quantum level. However, this is impractical due to the

© Springer Fachmedien Wiesbaden GmbH 2018
M. O. Steinhauser, *Multiscale Modeling and Simulation of Shock Wave-Induced Failure in Materials Science*,
https://doi.org/10.1007/978-3-658-21134-9_1

high cost of simulations and because properties of large ensembles of atoms are governed by complex physical phenomena.

This work is organized as follows: In Chap. 2 the theoretical foundations and some historical aspects of shock wave physics are briefly introduced and discussed.

In Chap. 3 the importance of multiscale modeling and simulation in science and technology is highlighted and discussed. We also provide a brief overview of the research codes which are used to derive the simulation results in this work. We end this chapter with the introduction of a new multiscale coupling method that allows for bridging the atomic and the macroscopic continuum domain. The efficiency of this method is shown with a shock tube simulation.

In Chap. 4 new ideas to generate micro-structures of granular materials in 3D based on two approaches are presented: One approach uses the mathematical theory of the power diagram (PD) which is an extension of the ordinary Voronoi diagram (VD) and allows for subsequent finite element (FE) analyses under shock loading conditions, as discussed in Sec. 4.1. The other approach is based on a mesh-free, i.e. particle-based, discrete element method (DEM), which is presented in Sec. 4.3. We also present in Sec. 4.2 accompanying high-speed impact experiments that serve as test cases and validation for the numerical tools.

In Chap. 5 we introduce coarse-graining in soft matter systems and proceed to show that with coarse-grained polymer models universal crossover scaling laws in the transition from fully flexible polymers to semi-flexible behavior can be recovered from simulations. This is in contrast to contradicting speculations in the literature about the possible scaling exponents based on atomistic models of semi-flexible polymers. We are able to show that these discussions were in vain since the atomistic models used in the literature are not able to simulate long enough chains to obtain the persistence lengths necessary to reveal the correct scaling laws. We then proceed with the introduction of a new coarse-grained model for phospholipid membranes and we investigate its equilibrium properties, determining a phase diagram of our membrane model including regions were fluid behavior dominates.

We also discuss solid-like membranes, and even completely closed vesicles.

In Chap. 6 we discuss experiments on the laser-induced shock wave destruction of human brain tumor cells. Here, we introduce our new and improved experimental setup for shock wave generation in tumor cells and corresponding pressure measurements. Additionally we use the technique of photon Doppler velocimetry (PDV) to generate very precise velocity data of the generated shock wave to be used as input for numerical simulations. We manage to figure out the shock pressure levels necessary to achieve a convincing destructive biological effect in the tumor cells and underpin our findings with data from cell viability measurements based on the Coulter principle, trypan blue treatment and a MTT viability test over five days and several cell cycles. Furthermore, we present shock wave simulations of large membranes, performed with the ingredients of the previously introduced multiscale coupling method and coarse-grained modeling. A quantitative analysis of membrane damage due to a shock wave is performed. We observe self-repair in the membrane after shock wave impact which has also been reported in experiments. We further find the existence of a limiting shock wave speed beyond which damage in a membrane becomes irreversible.

In Chap. 7 we end this work with a summary and draw some conclusions.

2. What are shock waves?

The theoretical foundations of shock wave[1] physics originated from 18th-century studies of classical acoustics and 19th-century aeroballistics. With the advent of high-speed photography in late 19th century and further progress in experimental visualization techniques in 20th century, the study of shock waves in different states of matter has gradually emerged from a very small and unnoticed branch of physics to a major complex and interdisciplinary science [129, 222]. Finally, with the sustained progress in ever faster computer hardware since late 20th century, and the increasing availability of computational resources, shock wave research – traditionally an experimental science – became more and more the object of computational investigations, where the non-linear equations of shock wave theory are solved numerically.

Shock waves are in essence discontinuous, rapid mechanical phenomena that have been observed and studied in the laboratory and in nature, in microscopic as well as in macroscopic dimensions and in all states of matter: gaseous [192, 257], liquid [32, 74, 111], solid [31, 116, 120], plasma [92, 93], and even in Bose-Einstein condensates [39, 52, 65]. Terrestrial examples [163] of naturally occurring discontinuities are high-energy events such as meteorite impacts, thunder, volcanic explosions, sea- and earthquakes or tsunamis, while in outer space [30, 158] they encompass plasma shock waves induced by solar wind, supernovae explosions, implosions of white dwarfs, comet and asteroid impacts, and stellar or galactic jets. In the laboratory, shock waves are produced by sudden release of energy like in explosions, supersonic flows or by the impact of projectiles at high speeds.

Essentially, shock waves exist because of the intrinsic non-linear nature of the hydrodynamic equations – which we shortly review in

---

[1]Sometimes also written as "shock-wave" or "shockwave".

© Springer Fachmedien Wiesbaden GmbH 2018
M. O. Steinhauser, *Multiscale Modeling and Simulation of Shock Wave-Induced Failure in Materials Science*,
https://doi.org/10.1007/978-3-658-21134-9_2

Sec. 2.2 – stemming from terms like $\vec{v}\vec{\nabla}\vec{v}$ in the equation of motion. Typically, the motion of fluids or gases are described on the basis of macroscopic fields in space-time such as velocity $\vec{v}(\vec{r}, t)$, pressure $p(\vec{r}, t)$, or density $\rho(\vec{r}, t)$. However, because of the fundamentally discrete, quantum-mechanical nature of matter, an exact theory of shock waves – i.e. a theory not based on the continuum approximation – would involve quantum mechanical concepts, unifying the notion of discrete matter particles with the one of continuous functions in space-time which denote physical properties such as particle or mass densities. A discussion of the involved concepts and problems can be found e.g. in [214], particularly in Chapter 8 therein. Such a unifying theory of the particle/wave (or discrete/continuum) concepts is currently not at hand, so one usually has to make do with developing a theory of shock waves based on the conservation equations of mass, momentum and energy, which are expressed in the hydrodynamic equations, see (2.12)-(2.14). At any rate, the enormous variety of phenomena and in particular the possible applications of shock waves in different areas such as physics, chemistry, biology, medicine and even engineering renders the scientific study of shock waves a rather fascinating subject.

2.1. Definition of shock waves

In the natural sciences, a shock wave describes a mechanical wave characterized by a surface discontinuity in which, within a narrow region, thermodynamic quantities such as pressure p, density ρ, particle velocity $\vec{v}$ or temperature T change abruptly. Shock waves occur in supersonic flows where the flow velocity exceeds the (adiabatic) sound speed. With non-steady flows, the surface discontinuity generally is not at rest. Its velocity however has nothing to do with the flow velocity, as the fluid particles, i.e. mass, may pass through this surface.

A different type of discontinuity is the so-called *contact discontinuity* that can occur at any flow speed. A contact discontinuity is a surface separating two fluids or gases with different physical properties. Unlike a shock, there is *no* flow of mass across a contact discontinuity. While a theory of shock waves with an infinitesimal jump condition can be developed *mathematically* in the framework of approximative continuum models of condensed matter, i.e. based on the hydrodynamic equations of ideal fluids or gases (see Sec. 2.2), in *physical* applications, shock wave theory is somewhat limited, because the width of a shock wave is always of finite size and because of the actually discrete atomic nature of condensed matter. It turns out that shock waves are essentially small regions were non-adiabatic, i.e. irreversible energy dissipation occurs and for which the shock wave is a mathematical idealization. The width of the shock wave – i.e. the size of the dissipative region – establishes itself according to the conservation laws of continuum theory. In very strong, high-amplitude shock waves, their width becomes so small that they are practically indistinguishable from the mathematical idealization of an infinitesimally small perturbation with jump conditions for thermodynamic variables. In solids, physical shock waves are mechanical waves of finite amplitudes and arise when condensed matter is subjected to a rapid compression. Phenomenologically, shock waves can be defined by several major distinctive properties as follows:

Defining features of shock waves:

- a pressure-dependent, supersonic velocity of propagation,

- the formation of a steep wave front with a sudden change of thermodynamic quantities,

- non-linear superposition properties (for reflection and interaction),

- a strong decrease of propagation velocity with increasing distance from the center of the origin for the case of non-planar shock waves.

Sometimes another criterion for a shock wave is listed in shock wave literature which is an extremely short rise time of the pressure within

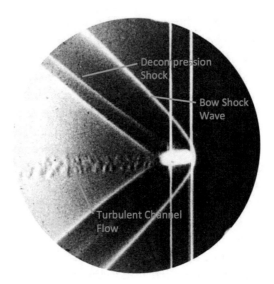

Figure 2.1.: Ernst Mach's Schlieren photograph of a bow shock wave
around a supersonic brass bullet (1887) [153]. This was one
of the first ever published photographs of a supersonic shock
wave in the history of shock wave research. The projectile
is moving from left to right in the figure. In general, one
distinguishes compression and decompression shock waves,
depending on whether the shock wave moves into a region
of lower or higher density, respectively.

tens of nanoseconds. However, this is not a defining criterion of
shock waves, because no generally agreed upon definition exists of
what "extremely short" really is supposed to mean – it is a term of
rather subjective nature. A nanosecond rise time certainly applies
to lithotripter shock waves which are used in medicine as a non-
invasive technique to comminute calculi [157] and for other therapeutic
purposes [147]; however, in the geologic realm for example, meteorite
impacts, explosive volcanic eruptions and earthquakes can provoke
drastic and irreversible changes within seconds.

One thing that distinguishes a shock wave from an ordinary sound
wave is that the initial disturbance in the medium that causes a shock

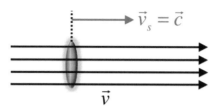

Figure 2.2.: Stationary flow of a fluid with velocity $\vec{v}$. A perturbation, e.g. a compression is shown which propagates relative to the rest system of the fluid with the velocity of the speed of sound $v_s = c$.

wave is always traveling at a velocity greater than the phase velocity of sound in the medium[2]. Because a shock wave moves faster than the speed of sound, the medium ahead of the shock front cannot respond until the shock strikes, and so the shock wave falling upon the particles of matter initially at rest, is a supersonic phenomenon.

Figure 2.1 exhibits one of the first ever published Schlieren picture of the bow shock wave generated by a projectile moving at supersonic speed. The two vertical lines visible in Fig. 2.1 are wires with a spark gap inserted for triggering the spark light source. The projectile had a speed of $v_0 \approx 530\,m/s$ which resulted in a Mach cone angle of $\alpha \approx 40°$ (cf. Fig. 2.2b). In a steady flow with constant velocity $\vec{v}$ the speed at which a perturbation travels relative to a reference frame in which the laboratory is at rest, is composed of two parts, see Figs. 2.2 and 2.3: On the one hand the perturbation travels with velocity $\vec{v}_s = \vec{c}$ relative to the fluid into a direction $\vec{n}$. On the other hand, the source of disturbance moves at the same time with the velocity $\vec{v}$ of the fluid flow, see Fig. 2.2. For simplicity, we assume here constant flow velocity $\vec{v}$ and consider a perturbation in the fluid at a fixed point O. This perturbation propagates with velocity $\vec{v} + c\vec{n}$ from O. We can see in Fig. 2.3 that the velocity depends on the direction of $\vec{n}$. Vector $\vec{n}$ is a

---

[2]The electromagnetic analog of this is called *Cherenkov radiation* [42], which is caused by a charged particle traveling through a dielectric medium at a velocity faster than the speed of light in that medium.

radial unit vector starting from point O with respect to the moving fluid flow. The possible values of the velocity lie on the surface of a sphere according to Fig. 2.3. If $v_s < c$, the vectors $\vec{v} + c\vec{n}$ can attain any direction in space, cf. Fig. 2.3 a). Hence, a flow that propagates with subsonic speed will affect the whole domain of the fluid. For $v_s > c$ the velocities $\vec{v} + c\vec{n}$ lie within a cone the tip of which is located in point O, cf. Fig. 2.3b. This cone is the envelope of the sphere, so we find for the total cone angle, the so-called *Mach angle* α the following relation:

Definition of Mach angle α and Mach number M_a:

The *Mach angle* is defined as:

$$\sin \alpha = \frac{c}{v}. \tag{2.1}$$

The ratio

$$M_a = \frac{v}{c} \tag{2.2}$$

is called *Mach number*. For $M_a > 1$, the flow is supersonic and for $M_a < 1$ it is subsonic, respectively.

Hence, we find that a perturbation travels only with supersonic speed in the direction of flow within the boundaries of a cone with angle 2α. Angle α gets smaller with increasing flow speed v, so the perturbation in a flow has no effect in the region outside that cone. The boundary layer of the region which can be influenced by a perturbation is called *Mach's cone, characteristic lines* or simply *characteristics*. The parts c) and d) in Fig. 2.3 exhibit the situation in a fluid at rest through which a perturbation moves with velocity $\vec{v}$.

Another difference between a shock wave and an ordinary sound wave is that the entropy is increased in a shock wave – hence, in contrast to ordinary sound waves, shock waves constitute an irreversible process. In engineering applications, shock waves produced in air by an explosion and radiating outward from its center are termed *blast*

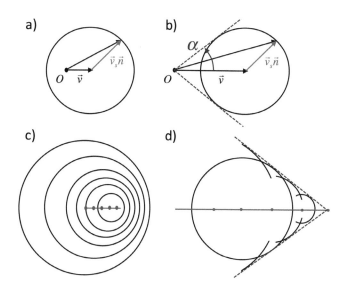

Figure 2.3.: Relative motion of a perturbation in a medium. A perturbation, e.g. a compression is shown which propagates relative
to the rest system of the fluid with the velocity of the speed
of sound $v_s = c$. a) Flow velocity is smaller than the velocity of sound, i.e. $v < c$. b) Flow velocity is larger than
the velocity of sound, i.e. $v > c$. c) Perturbation in a fluid
at rest. This could be e.g. a projectile with velocity $v_s < c$
moving from left to right in the figure. The circles in this
snapshot of the movement represent the positions of the
wave fronts that have been initiated at different positions
and points in time, represented by small dots along the
green line. There is *no* envelope of the individual waves. d)
The same as c) but with $v_s > c$. The individual waves have
a cone as common envelope.

waves, because they cause a strong wind, while the term shock wave
is preferred for such waves occurring in water or the ground, because
here the effect is like that of a sudden impact. Impact experiments are
often performed to study the fracture and failure behavior of solids
(see Chap 4) or in an attempt to determine a shock equation of state,
describing a material under extreme conditions of high pressure and/or

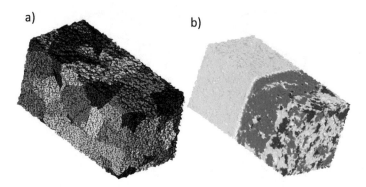

Figure 2.4.: Shock wave simulation in a material block of aluminum oxide (Al_2O_3) of dimension $(1.6 \times 1.6 \times 3.2)\mu m^3$. The model
consists of 176 grains with 1,500,000 finite elements (tetrahedra) in the analysis. a) A generated and optimized
structure with the method discussed in Chap. 4, indicating
the different grains in the material specimen in different colors. b) Snapshot of the shock wave simulation at 170ps after
initiation of the shock wave. The color code displays the
pressure levels in the range of $\pm 1 GPa$. Interestingly, typical for shock waves in solids, the shock front of the elastic
precursor wave (the very bright narrow shock front) advancing the plastic wave, can be detected. Adapted from [133].
© Martin O. Steinhauser.

temperature. Along with phenomenological constitutive equations,
this allows for numerically analyzing a material in a shocked state, e.g.
in a finite element analysis, cf. Fig. 2.4.

When a solid is compressed hydrostatically, its volume shrinks
and the pressure rises [234]. The hydrostatic curve is smooth as a
function of stress and volume. If the solid is compressed uniaxially,
stress and pressure also increase until the maximal shear stress of the
material is reached. Then it will yield and change from the uniaxially
compressed state to hydrostatic compression. Many solid materials
under moderate shock loading exhibit such a two-wave structure where
the first one is essentially an elastic wave followed by a slower plastic

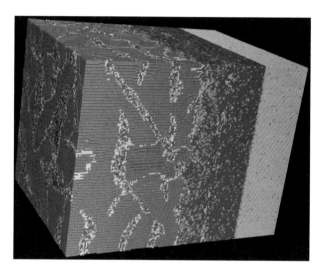

Figure 2.5.: Shock-induced phase transition in an atomic multi-million molecular dynamics simulation of a generic crystalline solid according to [117].

deformation wave front, which can be seen in Fig. 2.4, too. In the $p - V$-diagram this transition from elastic to plastic behavior is visible through a cusp, called the Hugoniot elastic limit (HEL), and there are a considerable amount of experimental data which illustrate these elastic–plastic–shock phenomena with an elastic precursor wave [11].

Very often, impact experiments with solids (discussed in Chap. 4) are used as test cases for the predictions obtained from numerical multiscale models – such as the one presented in Fig. 2.4 – which simulate a material specimen macroscopic in size but considering several microstructural features of the solid. Another important use of impact experiments in the laboratory is to study the crack initiation and propagation until ultimate failure occurs in a set-up that allows for very reproducible shock wave generation. One of these set-ups is the so-called *edge-on-impact test* where an impactor strikes the edge of a material specimen at high speed, initiating a well-defined plane shock wave through the material that ultimately leads to fracture and destruction (see Chap. 4). Also phase transitions in solids can be induced using shock waves and have been studied experimentally and numerically – see Fig. 2.5 – using classical molecular dynamics simulations in a variety of materials [41, 103, 104, 116, 117]. The largest MD simulation ever performed as yet was done at the Leibniz Supercomputing Centre in Munich in 2013, involving more than 4 trillion $(4, 125 \times 10^{12})$ particles interacting via a Lennard-Jones potential. This corresponds to a chunk of matter with an edge length of significantly more than $2.5 \mu m$, which was the system size of the previous record of 2008 [85].

2.2. The hydrodynamic equations

In this section we briefly review the origin of the hydrodynamic equations which are then used in Sec. 2.3 for the derivation of the Rankine-Hugoniot jump conditions at a shock wave front. The hydrodynamic equations for the description of continuous media such as elastic bodies, fluids and gases can be derived with mathematical rigor from the Liouville equation of the distribution function of a many particle system, but they can also be justified phenomenologically as arising from balance equations for mass, momentum and energy.

A system of N free particles can be approximated as a continuum when the mean free path length λ of individual particles (e.g. atoms or molecules) is negligible compared to the characteristic macroscopic linear dimension L of the system [136]. The concept of a fluid as being composed of *fluid elements* $\mathcal{F}$ of linear dimension $l_{\mathcal{F}}$ is useful if

$$\boxed{\lambda \ll l_{\mathcal{F}} \ll L.}$$ (2.3)

For the fluid elements, the physical properties – e.g. the density ρ, pressure p or the velocity of $\vec{v}$ a fluid element – are obtained in local thermal equilibrium as average values of the atomic/molecular quantities. Hence, the velocity $\vec{u}_p$ of of a particle

$$\vec{v}_p = \vec{v}_{\mathcal{F}} + \vec{\Gamma}_p$$ (2.4)

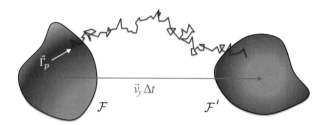

Figure 2.6.: In a fluid element $\mathcal{F}$ the velocity $\vec{v}_p$ of each particle is the sum of a stochastic component $\vec{\Gamma}_p$ and the average velocity $\vec{v}_{\mathcal{F}}$ of all particles. When the mean free path λ is small compared to the size of the fluid element, the latter is preserved.

is then composed of a statistical component $\vec{\Gamma}_p$ and the average velocity $\vec{v}_{\mathcal{F}}$ of the fluid element, see Fig. 2.6. As $\lambda \ll l_{\mathcal{F}}$, the fluid element is preserved except for a particle exchange along its boundary which may be described as a diffusion process. There are in essence two description methods that arise from the development of a continuum theory of matter.

Lagrangian description:

This is the perspective of an observer in a co–moving reference frame. Here, the dynamics of each individual fluid element is described at position $\vec{r}_0$ at time t_0, such that

$$\vec{v} = \frac{d\vec{r}}{dt}_{|\vec{r}=\vec{r}_0} \tag{2.5}$$

denotes the velocity of the fluid element. As a result one obtains the complete trajectory of each fluid element. This description cannot be sustained for very large systems ($N \gg 1$). In fact, most often one is not really interested in the dynamics of individual fluid elements but in the state of flow at a fixed location in space.

Eulerian description:

This is the most common perspective used in fluid dynamics, the perspective of an observer in a fixed laboratory system. In each point $\vec{r}$ one monitors the dynamics of the local, macroscopic observables, which are the degrees of freedom of this *field theory*. Hence, the fluid is represented by macroscopic physical fields, i.e. each point is space attains a certain scalar or vectorial value. Most often one is interested in:

- the pressure field $p(\vec{r}, t)$,

- the density field $\rho(\vec{r}, t)$, and

- the velocity field $\vec{v}(\vec{r}, r)$,

In the Eulerian description of fluids, $d\vec{r}/dt$ and $d^2\vec{r}/dt^2$ have no physical meaning as $\vec{r}$ is just a parameter (like time t) and no dynamic variable. The fluid flows at time t at position $\vec{r}$ with velocity $\vec{v}$ and the trajectories of individual fluid elements remain unknown. At location $\vec{r}$ one measures the velocities of different fluid elements at different times.

For the case

$$1 \gg \frac{hn^{1/3}}{\sqrt{3mk_BT}}, \tag{2.6}$$

where h is Planck's constant, n denotes particle density, m is particle mass, T is temperature, and k_B is Boltzmann's constant, the quantum mechanical wave packets of the individual particles do not overlap. Hence, the system is essentially an ensemble of N *classical* particles which – in principle – can be described by the $2N$ first order Hamiltonian equations of motion

$$\dot{\vec{r}}_i = \vec{\nabla}_{p_i} H, \tag{2.7}$$

$$\dot{\vec{p}}_i = -\vec{\nabla}_{r_i} H, \tag{2.8}$$

where $\vec{\nabla}_{p_i}$ and $\vec{\nabla}_{r_i}$ are a physicists' symbolic notation to denote the partial derivatives with respect to each of the three components of $\{r_i\}$ and $\{p_i\}$ with $i = 1, 2, 3$. The ensemble of N particles can also be described in its $6N$-dimensional phase space by using the *Liouville equation* for the probability distribution function ξ of this ensemble.

$$\partial_t \xi = \sum_{i=1}^{N} \left(\vec{\nabla}_{\vec{r}_i} H \cdot \vec{\nabla}_{\vec{p}_i} \xi - \vec{\nabla}_{\vec{p}_i} H \cdot \vec{\nabla}_{\vec{r}_i} \xi \right), \qquad (2.9)$$

where $\xi(\vec{r}_1, ..., \vec{r}_N, \vec{p}_1, ..., \vec{p}_N, t)\, d^3\vec{r}_1 \cdots d^3\vec{r}_N \cdots d^3\vec{p}_1 \cdots d^3\vec{p}_N$ denotes that part of the ensemble which can be found at time t in the volume element $d^3\vec{r}_1 \cdots d^3\vec{r}_N \cdots d^3\vec{p}_1 \cdots d^3\vec{p}_N$. The quantity ξ is observed, i.e. no particles of the ensemble are destroyed, nor created. For ensembles with large particle numbers ($N \gg 1$), a complete microscopic description is obviously not useful any more.

An approximative level of description of the N-particle system is given by introducing a statistical distribution function

$$f(\vec{r}, \vec{v}, t), \qquad (2.10)$$

where $f(\vec{r}, \vec{v}, t)\, d^3\vec{r}\, d^3\vec{v}$ is the probability of finding a particle at position $\vec{x}$ with velocity $\vec{v}$. With this function one can derive the so-called *hydrodynamic equations* [136]. Hence, the hydrodynamic equations of motion[3] are an approximative description of the dynamics of a many-particle system where one is not interested in the specific properties of the individual particles. The time evolution of f is given again by the Liouville equation, this time for the N-particle distribution function of the system which – for canonical ensembles – follows from *Liouville's theorem* [135].

By successive integration of the Liouville equation with respect to the coordinates of the N-particle distribution function one obtains a *non-linear, coupled integro-differential equation* of the form [135]

$$\partial_t f + \frac{\vec{v}}{m} \cdot \vec{\nabla}_{\vec{r}} f + \frac{\vec{F}}{m} \cdot \vec{\nabla}_{\vec{v}} f = (\partial_t f)_{\text{coll}}, \qquad (2.11)$$

---

[3]These are the Navier-Stokes and the Euler equations.

with an increasing order of particle correlations. The term on the right-hand side of (2.11) is the collisional term which is actually a functional of an angle-dependent elastic collision cross section of particles. The dynamics of the system can be described by a Boltzmann equation, i.e. as a dilute, neutral gas when the following conditions are met:

- the particle interactions are only short-ranged and independent of velocity,

- the effect of external forces (e.g. gravitation or Coulomb interaction) can be neglected during particle collisions,

- there are only two-particle collisions,

- there is molecular chaos, i.e. the local order that is established by a particle collision is completely destroyed before the next collision.

The evaluation of (2.11) is based on the definition of macroscopic properties as different moments of velocities and momenta and lead to *Maxwell-Boltzmann transport equations* [136]. The hydrodynamic equations follow from the transport equations for the first three lowest moments and an additional phenomenological equation of state $p = p(\rho)$:

The hydrodynamic equations:

$$\frac{\partial \rho}{\partial t} + \mathrm{div}(\rho \vec{v}) = 0, \qquad (2.12)$$

$$\frac{\partial (\rho \vec{v})}{\partial t} + \mathrm{div}(\rho \vec{v} \vec{v}) + \mathrm{grad}\, p = 0, \qquad (2.13)$$

$$\frac{\partial (\rho e)}{\partial t} + \mathrm{div}\left[(\rho e + p)\, \vec{v}\right] = 0. \qquad (2.14)$$

The hydrodynamic equations express conservation of particle number, respectively mass (2.12), conservation of momentum (2.13) and conservation of energy (2.14). One of the main difficulties of hydrodynamics is its intrinsic non-linearity, explicitly visible in the term $\text{div}(\rho\,\vec{v}\,\vec{v})$ in the above equation of motion. This makes it difficult to find exact solutions, except in those cases where there is a lot of symmetry. Another way to simplify, that is, to linearize the equations is to look at small perturbations around an equilibrium where there is a force balance. The equilibrium state is then a solution of the above equations and one looks at small deviations from that equilibrium, assuming that the changes in velocity $\vec{v}$, density ρ and pressure p remain small. If that is the case, nonlinear terms can be neglected when describing the evolution of these small perturbations and all variations in fluid quantities such as velocity, density and pressure can be expressed as *linear* functions of some displacement field $\xi(\vec{r},t)$ which describes how far individual fluid elements are displaced form their equilibrium position. This linearization technique works well as long as the amplitude of the displacement vector remains sufficiently small (which is not the case with shock waves). For viscous, self-gravitating fluids one obtains from the conservation of momentum the *Navier-Stokes-Equation*:

$$\frac{\partial}{\partial t}(\rho v_i) + \sum_{k=1}^{3} \frac{\partial}{\partial x_k}(\rho v_i v_k) + \sum_{k=1}^{3} \frac{\partial}{\partial x_k}\Pi_{ik} = -\rho(grad\,\Phi)_i\,, \qquad (2.15)$$

where Φ is the gravitational potential[4] and Π_{ik} are the components of the *momentum flux tensor* which is given by

$$\Pi_{ik} = p\delta_{ik} + \rho v_i v_k - \pi_{ik}\,, \qquad (2.16)$$

with the viscosity term

$$\pi_{ik} = \eta\left(\frac{\partial v_i}{\partial x_k} + \frac{\partial v_k}{\partial x_i} - \frac{2}{3}\delta_{ik}\text{div}\,\vec{v}\right) - \zeta\delta_{ik}\text{div}\,\vec{v}\,. \qquad (2.17)$$

The constants $\eta > 0$ and $\zeta > 0$ are called coefficients of viscosity. Note that the first term on the right hand side of (2.17) is trace-free, i.e. in the case of an incompressible fluid ($\text{div}\,\vec{v} = 0$), the viscosity is

---

[4]Defined via the Poisson equation $\Delta\Phi = 4\pi G\rho$.

only described by η. Coefficient η is often denoted *dynamic viscosity* and the ratio $\nu = \eta/\rho$ is called *kinematic viscosity*. The relative importance of inertia versus viscosity is determined by the *Reynolds number*

$$Re = \frac{vL}{\nu}\,. \tag{2.18}$$

One defining feature of a shock wave listed in Sec. 2.1 is a *supersonic* propagation of the perturbation. In this case, the Reynolds number will be large, i.e. $Re \gg 1$ and the viscosity has an important effect on the motion of the fluid only in a very small region. The motion is mostly dominated by inertia – the flow is turbulent and non-linear effects due to the term $\rho v_i v_k$ in (2.15) dominate. Hence, the theory of shock waves in continuous media is developed with the assumption of treating the medium as an *ideal* fluid. For an ideal fluid (neglecting viscosity and heat diffusion), the *Euler equation* holds:

$$\frac{\partial}{\partial t}(\rho v_i) + \sum_{k=1}^{3} \frac{\partial}{\partial x_k}(\rho v_i v_k) + (\text{grad}\,p)_i = -\rho(grad\,\Phi)_i\,, \tag{2.19}$$

2.3. Discontinuity surfaces

In a continuum description of condensed matter one assumes that all macroscopic quantities (velocity, pressure, density, etc.) are continuous functions. However, non-linear flows are possible in which discontinuities occur across one or more surfaces. Albeit the differential form of the hydrodynamic equations allows for a complete description of continuous flows, such discontinuities can only be described by using the integral form of the hydrodynamic equations. When using their differential form, the equations have to be supplemented by so-called "jump conditions" for the discontinuities. In non-steady flows such discontinuity surfaces do not remain fixed with respect to the velocity of the fluid particles. Hence, fluid particles may cross a surface of discontinuity in their motion. To formulate the *jump conditions* we consider an element of the surface $d\vec{F}$ and use a reference frame

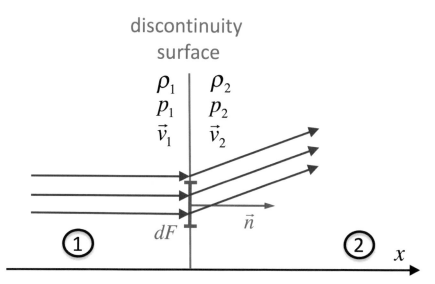

Figure 2.7.: Geometry considered for deriving the jump conditions at a discontinuity surface in a coordinate system Σ which is at rest relative to an observer that moves with the discontinuity. We consider a small surface element $d\vec{F} = dF \cdot \vec{n}$ with normal vector $\vec{n}$. The three arrows crossing the discontinuity surface symbolize the jump conditions for the thermodynamic quantities.

Σ which is fixed to this element, i.e. we take the point of view of an observer in a *co–moving reference frame*. We further assume for simplicity, that the flow is one-dimensional along the x-axis which is oriented along the surface normal $\vec{n}$, see Fig. 2.7. There shall be no external forces. We then find the following three conditions, based on the hydrodynamic conservation equations:

1. **The mass flux through the surface must be continuous.**
   There is no mass created or destroyed. Hence, the mass flux $j = \rho v$ (with $v = v_x = \vec{v} \cdot \vec{e}_x$) at the two sides of the surface is given by:

$$\rho_1 v_{1x} = \rho_2 v_{2x}, \tag{2.20}$$

where the indices 1 and 2 indicate the values of any quantity on the two sides of the surface as indicated in Fig. 2.7.

To simplify things, we want to use the following abbreviation:

Definititon of the discontinuity bracket [ ]:

We use the square brackets to denote the difference between the values of any quantity A on the two sides (denoted as "1" and "2") of the discontinuity surface, cf. Fig. 2.7:

1. For scalars:
$$[A] = A_1 - A_2 \tag{2.21}$$

2. For vectors, the bracket is understood to be taken for each Cartesian component separately:
$$[\vec{A}] = \begin{cases} A_{x1} - A_{x2} & , \\ A_{y1} - A_{y2} & , \\ A_{z1} - A_{z2} & . \end{cases} \tag{2.22}$$

Example:
$$[B\vec{A}] = \begin{cases} B_1 A_{x1} - B_2 A_{x2} & , \\ B_1 A_{y1} - B_2 A_{y2} & , \\ B_1 A_{z1} - B_2 A_{z2} & . \end{cases}$$

Using Eq. (2.21) we obtain as our first jump condition:

$$\boxed{[\rho v_x] = 0.} \tag{2.23}$$

The y- and z-components of the mass flux, i.e. the flux parallel to the discontinuity surface are not subject to any restrictions.

2. **The energy flux through the surface must be continuous.** The energy flux in continuum theory is given by $\rho \vec{v} \left(\frac{\vec{v}^2}{2} + e + \frac{p}{\rho} \right)$ where e is the specific internal energy. Hence, we have the condition

$$\left[\rho v_x \left(\frac{v^2}{2} + h \right) \right] = 0, \tag{2.24}$$

where we have introduced the *specific enthalpy*

$$h = e + \frac{p}{\rho}. \tag{2.25}$$

3. **The momentum flux through the surface must be continuous.** The momentum flux tensor Π_{ik} in continuum theory is given by (2.16). Hence, when neglecting viscosity effects, the flux of the i-th momentum component per unit area, orthogonal to the x_k-axis is given by

$$\Pi_{ik} n_k = p n_i + \rho v_i v_k n_k = \vec{p}\,\vec{n} + \rho \vec{v} \left(\vec{v} \cdot \vec{n} \right).$$

Due to our particular choice of the coordinate system with $\vec{n} = \vec{e}_x$ we finally have three momentum jump conditions which we can succinctly summarize in vector form with (2.21) and (2.22):

$$\left[\rho v_x \vec{v} + p \vec{n} \right] = 0. \tag{2.26}$$

2.3.1. Rankine-Hugoniot jump conditions

Equations (2.23)–(2.26) form a complete system of boundary conditions at a discontinuity surface. This allows for deducing *two types of discontinuity surfaces.*

- The *first type* is characterized by zero mass flux through the surface, i.e. $\rho_1 v_{x1} = \rho_2 v_{x2} = 0$. As ρ_1 and ρ_2 are both not zero, it follows that $v_{x1} = v_{x2} = 0$. Inserting this result into Eq. (2.26) yields the condition $p_1 = p_2$. Hence, the normal velocity component and the pressure are continuous at the discontinuity surface:

$$v_{x1} = v_{x2} = 0, \quad [p] = 0. \tag{2.27}$$

On the other hand, the tangential velocity components v_y and v_z along with all other thermodynamic quantities (except for the presssure) may be discontinuous by any amount. This kind of discontinuity is called a *tangential discontinuity*.

- The *second type* is characterized by mass flowing through the discontinuity surface. Thus, v_{x1} and v_{x2} are not zero and from combining (2.23) and (2.26) we obtain

$$[v_y] = 0, \quad [v_z] = 0. \tag{2.28}$$

Hence, with this type of discontinuity surface, the tangential velocity components are continuous, but the normal component has a jump as well as pressure and density. In (2.24) we can cancel ρv_x by (2.23) and replace v^2 by $v_x{}^2$, since v_y and v_z are continuous. Discontinuities of this type, i.e. which exhibit a mass flux through the surface, are called *shock waves*.

The term "shock wave" is actually quite delusive as it suggests that something is periodically oscillating, like in a wave. Yet, this term denotes just the final stage of a nonlinearly steepening wave – a *discontinuity surface*[5] – that has reached a balance between steepening and energy dissipation. Thus, the *jump conditions in shock waves* are given by the following *Rankine-Hugoniot equations*:

---

[5]Sometimes this surface is also called *shock wave front*, *shock front*, or simply *shock*.

Rankine-Hugoniot jump conditions for shock waves in a co-moving reference frame:

$$[\rho v_x] = 0. \tag{2.29}$$

$$\left[\frac{v_x{}^2}{2} + h\right] = 0, \tag{2.30}$$

$$[\rho v_x{}^2 + p] = 0, \tag{2.31}$$

When one returns to a coordinate system Σ' that is not co–moving with the discontinuity, see Fig. 2.8, one has to replace all occurrences of v_x in the Rankine-Hugoniot equations (2.29)–(2.31) by the difference between the velocity component of the flow v_n' normal to the discontinuity surface and the velocity $\vec{v}_s'$ of the surface itself, which – by our choice of geometry – is normal to the surface as well. Note, that the primed quantities are all measured in the laboratory system Σ'. From Fig. 2.8 we can see that

$$\vec{v}' = v_s' \cdot \vec{n} + \vec{v}, \tag{2.32}$$

where $\vec{v}$ is the velocity of the flow field measured in the rest frame Σ of the surface element $d\vec{F} = dF \cdot \vec{n}$. From (2.32) it follows

$$v_x = \vec{v} \cdot \vec{n} = \vec{v}' \cdot \vec{n} - v_s'. \tag{2.33}$$

In the following equation we introduce the quantity U as an abbreviation for the difference in velocities of the flow field and the shock discontinuity:

$$\boxed{U = \vec{v}' \cdot \vec{n} - v_s'.} \tag{2.34}$$

Note that U is determined by quantities that are measured in the laboratory system Σ'. However, to simplify our notation, we drop here the primed index with U. Hence, we transform from the rest system Σ of the discontinuity surface to the laboratory system Σ' by replacing

v_x with U in Eqs. (2.29)–(2.31) to obtain the Rankine-Hugoniot equations in the form:

Rankine-Hugoniot equations in the laboratory system:

$$[\rho U] = 0. \tag{2.35}$$

$$\left[\frac{U^2}{2} + h\right] = 0, \tag{2.36}$$

$$[\rho U^2 + p] = 0, \tag{2.37}$$

The Rankine-Hugoniot equations are a set of 3 equations with 8 unknowns $(v_{1/2}, \rho_{1/2}, p_{1/2}, h_{1/2})$. Two more equations that can be consulted are the

- Thermodynamic equation of state: A functional relation between thermodynamic variables. For example: $pV = nRT$ (classical ideal gas).

- Caloric equation of state: A temperature dependence of internal energy or heat capacity. For example: $h_2 - h_1 = c_p(T_2 - T_1)$.

Thus, three equations for six unknowns remain. Experimentally, temperature and pressure in front of the shock wave, i.e. T_1 and p_1 can easily be measured. If one more variable can be measured, the system of equations (2.29)–(2.31), respectively (2.35)–(2.37) becomes solvable. Usually, of the remaining unknown quantities most readily available is the velocity v_1 of the shock wave which can be measured, e.g. with hydrophones or other appropriate pressure sensors.

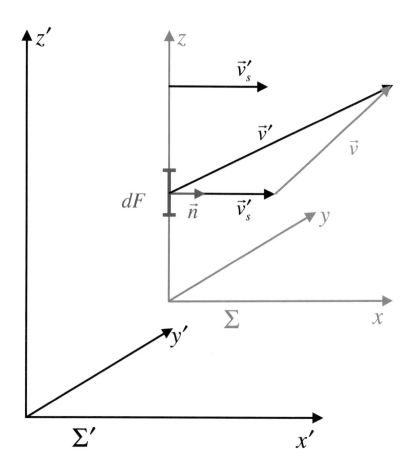

Figure 2.8.: Reference frames for the Galilei-transformation of the Rankine-Hugoniot equations. The hydrodynamic equations are Galilean-invariant. Thus one can transform from the description of shock waves in a co-moving reference frame Σ which is at rest in the system of the moving discontinuity surface to a laboratory system Σ'. The quantities calculated in the two frames are color-coded appropriately. Note that the surface element dF and the unit normal vector $\vec{n}$ are both invariant under Galilei-transformations.

2.4. Steepening of sound waves and Riemann characteristics

The wave equation describing changes in the density of sound waves is given by:

$$\frac{\partial^2 \rho}{\partial t^2} = c_s{}^2 \nabla^2 \rho, \tag{2.38}$$

with the velocity of sound

$$c_s{}^2 = \left(\frac{dp}{d\rho}\right). \tag{2.39}$$

The general solution of (2.38) is any function ρ of the form:

$$\rho(\vec{r}, t) = f(\vec{r} + c_s t) + g(\vec{r} - c_s t), \tag{2.40}$$

which describes the propagation of a plane wave front as a disturbance in negative (f) or positive (g) $\vec{r}$ direction. Figure 2.9 shows the situation for a one-dimensional wave where a given initial density profile moves unaltered in positive x direction. The distance covered by the wave is given by ($x_1 - x_0$) and the wave speed is the velocity of sound in the medium $c_s = (x_1 - x_0)/\Delta t$. However, for adiabatic changes of state, the velocity of sound c_s^{ad} is a function of density.

The origin of a theoretical treatment of shock waves, i.e. of discontinuities that may arise due to the non-linearity of the hydrodynamic equations, is Bernhard Riemann's theory of the propagation of acoustic disturbances of 1860 („Über die Fortplanzung ebener Luftwellen endlicher Schwingsweite") [191]. This paper is comparatively easy to read as it uses a notation very similar to that used today. Early in the paper, Riemann introduces what is known today as *Riemann variables* which he denotes r and s. Shortly after introducing these variables, Riemann describes how a compression wave would necessarily steepen leading to multiple values of the density ρ at one point. Then he continues[6]:

---

[6]My own translation from the German original in [191].

"Now, since this cannot occur in reality, then a circumstance would have to occur which renders this law invalid. [...] and from this moment on a discontinuity occurs, such that a larger value of ρ will directly follow a smaller one. The compression waves [Verdichtungswellen], that is, the portions of the wave where the density decreases in the direction of propagation, will accordingly become increasingly more narrow as it progresses, and finally go over into compression shocks [Verdichtungsstösse]".

He derives the jumps in mass and momentum for an isentropic (reversible) flow and establishes that the speed of a shock wave v_s is bounded by

$$v_1 + \sqrt{(\varphi'(\rho_1))} > v_s > v_2 + \sqrt{(\varphi'(\rho_2))},$$

where φ is some continuous function. He then continues to discuss what is known today as the *Riemann problem*, i.e. the wave patterns corresponding to various initial conditions with jumps in velocity u and ρ at some point x in space.

The Riemann problem:

In the theory of hyperbolic equations, if the state of a system is given by a function

$$\vec{U}\big(\rho(x,t), v(x,t), e(x,t)\big), \tag{2.41}$$

one speaks of a Riemann problem if the initial state $(t = 0)$ of the system is characterized by a jump condition:

$$\vec{U}(x,0) = \begin{cases} \vec{U}_l & x \leq 0 \\ \vec{U}_r & x > 0 \end{cases}, \tag{2.42}$$

where $\vec{U}_{l,r} = (\rho_{l,r}, v_{l,r}, e_{l,r})$ are the constant initial conditions at the left and the right side of $(x = 0)$.

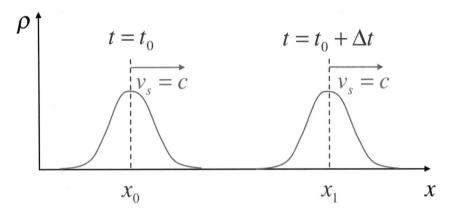

Figure 2.9.: Propagation of a one-dimensional wave. The initial density
profile of the wave propagates unchanged with time.

The special case $U_{l,r}(x,0) = 0$ is called *shock tube* and for hydrodynamic problems such *Sod shock tube tests* are carried out in laboratory experiments [145] to test predictions of numerical hydrodynamics
algorithms and the results of computer codes. The first such (theoretical) test was done by Sod in 1978 [204].

The Riemann problem can be solved with the *method of characteristics*. The solutions, the *Riemann invariants* $R_\pm$ turn out to be constant
along straight lines, the *characteristic curves* $X_\pm$, cf. Fig. 2.10.

$$R_+ = v_s + \frac{2c}{\gamma - 1} \tag{2.43}$$

is constant along the characteristic line $X_+(t)$ with

$$\frac{dX_+}{dt} = v_s + c = const. \tag{2.44}$$

and

$$R_- = v_s - \frac{2c}{\gamma - 1} \tag{2.45}$$

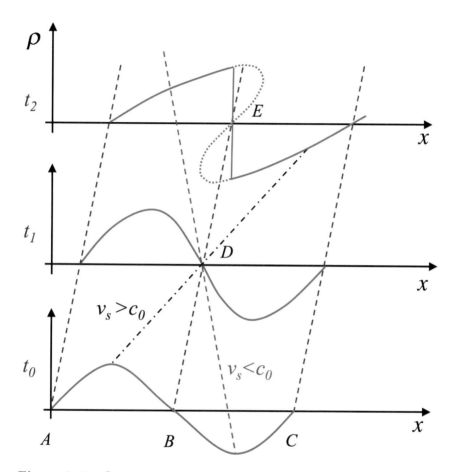

Figure 2.10.: Spacetime diagram of the steepening of an initially linear
sound wave to form a shock wave. Characteristic lines
pertaining to different wave propagation velocities are displayed. The profile of the wave changes with time. For the
case t_2 the solution would be ambiguous. Hence, instead of
the ambiguous wave profile (dotted curve), a shock wave
profile occurs, as displayed with the straight line (crossing
point E) which indicates a surface of discontinuity.

is constant along the characteristic line $X_-(t)$ with

$$\frac{dX_+}{dt} = v_s - c = const. \tag{2.46}$$

In Fig. 2.10 the density profile of a wave is shown in a spacetime diagram at three different points in time. At the beginning ($t = t_0$), the density at points A, B and C is the same, say ρ_0. The velocity of sound in these three points is also identical, i.e. disturbances from these points propagate with the same speed c_0 (blue/dashed parallel lines). In regions where $\rho > \rho_0$ (between A and B) the velocity of sound is larger than c_0, i.e. $v_s > c_0$. This means that the disturbance propagates faster (black/dot-dashed line crossing point D). At the same time, between B and C we have $c_s < c_0$ and a slower propagation of a disturbance (green/dashed line crossing point D). So, with non-linear equations, different characteristics may have different slopes and cross each other. At crossing points (e.g. point D in Fig. 2.10), the solution functions are not bijective anymore. Such functions are actually unphysical, which gives rise to a shock wave. This means that different parts of the shock wave profile propagate with different velocities which is called steepening of a wave: The wave crest travels faster than the wave trough and passes it with supersonic speed $v_s > c_0$, see the shock wave profile at $t = t_2$ in Fig. 2.10. Further steepening in the time evolution is usually stopped via viscous forces, see Fig. 2.11. The larger the viscosity η of the medium, the larger the width of the resulting (physical) shock wave. An ideal (mathematical) shock wave is obtained for extremely large shock amplitudes and for undisturbed media (when η is small or zero).

When perturbations with large amplitudes occur, then they will propagate faster in regions with higher density. As a consequence, changes in the shape of the perturbation occur. Generally speaking, waves can only propagate without change of shape, when the governing differential equations are *linear*. The hydrodynamic equations however, are non-linear and thus steepening of waves occurs: shock waves arise.

In Fig. 2.12 a) we exhibit a Sod shock tube, where a piston is moved with velocity v_p into a solid, a gas or a fluid, thus compressing

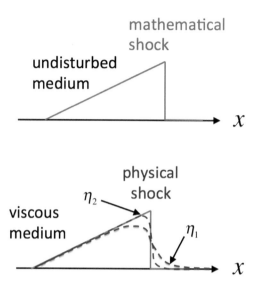

Figure 2.11.: A shock wave in undisturbed/viscous media. Top: the ideal, mathematical case, without viscosity. Bottom: The case of a physical shock wave, where the viscosity η (here: $\eta_1 > \eta_2$) increases the width of a shock wave. For zero viscosity one can assume a mathematical, ideal shock wave with infinitesimal thickness.

the material. We use this principle of generating a shock wave in our multiscale simulations of granular materials (Chap. 4) and of coarse-grained multiscale models of biomembranes as discussed in Chap. 6.

The interest in studying the shock tube problem is threefold. From a fundamental point of view, it offers an interesting framework to introduce some basic notions about nonlinear hyperbolic systems of partial differential equations. From a numerical point of view, this problem constitutes, since the exact solution is known, an inevitable and difficult test case for any numerical method dealing with discontinuous solutions. Finally, there is a practical interest, since this model is used to describe real shock tube experimental devices, sometimes

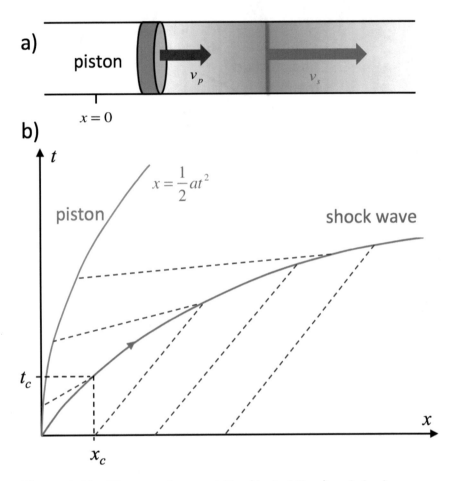

Figure 2.12.: Riemann characteristics (dashed lines) and shock wave
with a uniformly accelerated piston in a shock tube.

also called "gas guns", in particular, when the shock wave pressure is
used to accelerate small masses. The shock front has velocity v_s and
propagates at a speed larger than v_p into the material. In Fig. 2.12 b)
we show the corresponding spacetime diagram of the shock tube for
the case that the piston is moved with constant acceleration a. Thus,
we have $x = \frac{1}{2} at^2$. At $t = 0$ the piston is at $x(0) = 0$. The material is

at $x \geq \frac{1}{2}at^2$. When the piston moves, the temperature in the material increases upon compression. Since the velocity of sound in the material $c_0 \propto \sqrt{T}$ one can easily see, that the Riemann characteristics that were created at an earlier time are passed by ones that were created at a later point in time.

The Riemann invariants R_- (constant along the X_- characteristics) that are created at point $t = 0$ are given by

$$R_- = v_s \frac{2c}{\gamma - 1} = \frac{2c_0}{\gamma - 1}. \tag{2.47}$$

Hence, it follows:

$$c = c_0 + \frac{1}{2}v_s(\gamma - 1). \tag{2.48}$$

For the X_+ characteristics, which are created at the piston we have:

$$\frac{dX_+}{dt} = v_s + c = c_0 + \frac{1}{2}v_s(\gamma + 1)at_0. \tag{2.49}$$

From this it follows:

$$X_+(t, t_0) = \frac{1}{2}at_0^2 + \left(c_0 + \frac{1}{2}(\gamma + 1)at_0\right)(t - t_0). \tag{2.50}$$

The slope of characteristics increases with increasing t_0. This means, two characteristics cross each other at a finite point in time t_c. Let's assume that neighboring characteristics intersect first, i.e. $\delta t_0 \ll 1$. We then have:

$$X_+(t, t_0 + \delta t_0) \approx X_+(t, t_0) + \delta t_0 \frac{\partial X_+}{\partial t}(t, t_0), \tag{2.51}$$

i.e. neighboring characteristics intersect, when

$$\frac{\partial X_+}{\partial t} = 0. \tag{2.52}$$

From the preceeding equations we thus have:

$$t = \frac{2c_0}{a(\gamma + 1)} + \frac{2\gamma}{\gamma + 1}t_0. \tag{2.53}$$

This occurs for the first time with the characteristic which is created at $t = t_0$. Hence

$$t_c = \frac{2c_0}{a(\gamma + 1)} \quad \text{and} \quad x_c = \frac{2c_0{}^2}{a(\gamma + 1)}. \quad (2.54)$$

For example, assume $v_s = 100\text{ms}^{-1}$, $\gamma_{\text{air}} \approx 1.4$. We then have $t_c \approx$ 2.84s and $x_c \approx 969$m.

Generally speaking, the brilliant mathematician Riemann (and also Sir George G. Stokes before him [249]), did not fully grasp the true nature of the shock layer because it is actually a problem of *physics and thermodynamics*. One problem at that time was the lack of foundations in thermodynamics and in mathematics on the theory of generalized solutions to hyperbolic partial differential equations, where the first steps were not taken until the beginning 20th century [151]. The mathematical theory matured finally in the works of Sobolev [203] and Schwartz [198] treating discontinuities in the framework of distributions (often called δ–functions).

After Riemann's work, it was not until the works of Rankine in 1870 and later Hugoniot in 1887 that a full physical theory of shock discontinuities was established. Rankine was the first to show that within a shock a non-adiabatic process must occur, where the particles exchange heat with each other, but no heat is received from the outside [188]. In his work he goes on to find, for a perfect gas, the jump conditions for a shock wave moving with speed v_s into an undisturbed medium with pressure p and specific volume

$$V = \frac{1}{\rho}. \quad (2.55)$$

Hugoniot published his memoirs in two parts. The first one was published in 1887 [110] and consists of three chapters. The first chapter begins with an exposition of the theory of characteristic curves for partial differential equations of which he says:

> "The theories set out herein are not entirely new; however,
> they are currently being expounded in the works of Monge

and Ampère and have not, to my knowledge, been brought together to form a body of policy".

In the second chapter he provides the equations of motion for a perfect gas and then discusses the motion in gases in the absence of discontinuities in the third chapter. It is really mostly the second memoir of 1889 [110] that is interesting. Here, in chapter four Hugoniot treats the motion of a non-conducting fluid in the absence of external forces, friction and viscosity, and he shows that conservation of energy implies conservation of entropy in smooth regions and a jump of entropy across a shock. In chapter 5 he discusses the phenomena when jumps, i.e. discontinuities are introduced into the motion. In this last chapter, Hugoniot writes down his famous *Hugoniot equation*, which establishes which states are possible across a shock wave. It is written in the form

$$\frac{p + p_1}{2} = \frac{p_1 - p}{(m-1)(z_1 - 1)} + \frac{p_1 z_1 - pz}{(m-1)(z_1 - z)}, \qquad (2.56)$$

with $m = \gamma$, the ratio of specific heats, $V = z + 1$, and $e = pV/(\gamma - 1)$. Equation (2.56) can be used to derive this relation in the usually stated form as in (2.68), which connects the three thermodynamic variables internal energy e, pressure p and volume V in a rather elegant way. Note that in (2.68), which is sometimes called *Hugoniot shock equation* or *Hugoniot function*, the change of internal energy does not depend on the shock wave velocity. Actually, (2.68) allows to define a *shock adiabatic* in a p-V-diagram, which we will discuss in the next section.

2.5. Change of thermodynamic variables across shock waves

If pressure, density and velocity are known at both sides of a shock, then by using the set of equations (2.29)–(2.31) one can derive the change of thermodynamic variables across a shock wave. Let's consider again a coordinate system at rest in the rest frame of the shock wave, where the normal component of the discontinuity points in x-direction

(Fig. 2.7). Then we can simplify the notation in (2.29)–(2.31) by substituting v_x by simply v. We then have

$$\rho_1 v_1 = \rho_2 v_2 = j, \tag{2.57}$$

$$h_1 + \frac{v_1{}^2}{2} = h_2 + \frac{v_2{}^2}{2}, \tag{2.58}$$

$$p_1 + \rho_1 v_1{}^2 = p_2 + \rho_2 v_2{}^2, \tag{2.59}$$

where we denote the mass flux density at the discontinuity surface with $\vec{j} = j\vec{n}$. Using the specific volumes V_1 and V_2 according to (2.55), we can derive from (2.57):

$$v_1 = jV_1, \quad \text{and} \quad v_2 = jV_2. \tag{2.60}$$

Inserting (2.60) into (2.59) we find

$$\boxed{j^2 = \frac{p_2 - p_1}{V_1 - V_2} = \frac{p_2 - p_1}{\rho_2 - \rho_1}\rho_1\rho_2.} \tag{2.61}$$

On the right hand side of (2.61) we have substituted again the specific volumes V_1 and V_2 with the densities according to (2.55).

Equation (2.61) relates the propagation of a shock wave to the pressures and volumes (respectively densities) on the two sides of the discontinuity surface. From (2.61) and j always being positive we deduce that we always have:

$$p_2 > p_1 \quad \text{and} \quad V_1 > V_2, \tag{2.62}$$

$$p_2 < p_1 \quad \text{and} \quad V_1 < V_2, \tag{2.63}$$

For the velocity difference we obtain finally:

$$\boxed{v_1 - v_2 = \sqrt{(p_2 - p_1)(V_1 - V_2)}.} \tag{2.64}$$

Next we insert (2.60) into (2.58) to obtain

$$h_1 + \frac{j^2 V_1{}^2}{2} = h_2 + \frac{j^2 V_2{}^2}{2}, \tag{2.65}$$

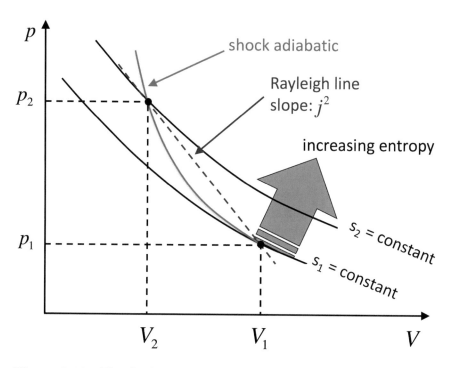

Figure 2.13.: The shock adiabatic which allows for determining the possible post-shock states (p_2, V_2) with given pre-shock state (p_1, V_1). The two dark (respectively black) curves are normal adiabatic curves of constant entropy. After the shock the system has higher specific entropy s, i.e. $s_2 > s_1$.

or, more succinct, using our above definition of the discontinuity bracket:

$$[h] = -\frac{j^2}{2}\left[V^2\right].\tag{2.66}$$

Substituting j^2 in (2.65) by the left equation of (2.61) we obtain

$$\boxed{h_2 - h_1 = \frac{1}{2}(V_1 + V_2)(p_2 - p_1)\,.}\tag{2.67}$$

Using the specific internal energy $e = h - pV$ in (2.67) we obtain

$$e_2 - e_1 = \frac{1}{2}(V_1 - V_2)(p_1 + p_2) \,. \qquad (2.68)$$

Equations (2.67) and (2.68) connect the thermodynamic variables p and V with e and h, respectively, on the two sides of a shock wave. On the other hand, these quantities are also connected by an equation of state, so for given values of p_1 and V_1 the above equations relate these values with p_2 and V_2. This relation is called *Hugoniot adiabatic* or *shock adiabatic* and is represented in Fig. 2.13. The shock adiabatic has only *one* intersection with the point (p_1, V_1) and likewise for (p_2, V_2). This means that the shock adiabatic is separated by the line $V = V_1$ into two parts, each of which lies entirely on one side of the line. The shock adiabatic is not (like the Poisson adiabatic) a continuous function that can be written in the form $f(p, V) = \text{const}$, where f is some function. Instead, it is determined by two parameters, the initial values p_1 and V_1.

According to the laws of thermodynamics, the entropy – and of course, also the specific entropy s has to increase during the motion of a flow. Thus, after crossing a shock wave, the specific entropy s_2 has to be larger than the initial value s_1:

$$s_2 > s_1 \,. \qquad (2.69)$$

An increase of entropy means that this flow is irreversible, which is connected with energy dissipation. It can be shown via Taylor expansion of (2.67) that the change in entropy $\delta s = s_2 - s_1$ in a shock wave is of third order of smallness with respect to the change of pressure $\delta p = p_2 - p_1$:

$$\delta s = \frac{1}{12\, T_1}\left(\frac{\partial^2 V}{\partial p_1{}^2}\right)(\delta p)^3, \qquad (2.70)$$

where T_1 is the temperature in point (p_1, V_1). The adiabatic compressibility $(\partial V/\partial p)_{|s}$ decreases with increasing pressure and the second derivative

$$(\partial^2 V/\partial p^2)_{|s} > 0 \,, \qquad (2.71)$$

so the pressure increases upon crossing a shock wave, i.e.

$$\boxed{p_2 > p_1.}$$
(2.72)

Considering (2.61) again in the form

$$j^2 = -\frac{p_2 - p_1}{\frac{1}{\rho_2} - \frac{1}{\rho_1}}.$$
(2.73)

shows that the squared mass flux density at the shock wave equals the negative slope of the Rayleigh line in point (p_1, V_1) which is given by:

$$\frac{\partial p}{\partial V} = -\rho^2 \frac{\partial p}{\partial \rho}$$
(2.74)

Using (2.74) with (2.73) and considering from Fig. 2.13 that in point (p_1, V_1) the slope of the Rayleigh line is larger than the slope of the tangent, we obtain

$$j^2 = \rho^2 \frac{\partial p}{\partial \rho}.$$
(2.75)

Equivalently – returning to the velocities and assuming a coordinate system Σ in the rest frame of the shock wave – we can write:

$$v_1{}^2 > \frac{\partial p}{\partial \rho}.$$
(2.76)

In (2.70) we saw, that δs is of third order in smallness with respect to δp. This means that we can approximately assume s to be constant in the vicinity of point (p_1, V_1). As a consequence we have

$$v_1{}^2 > \left(\frac{\partial p}{\partial \rho}\right) = c_1{}^2,$$
(2.77)

where $c_1{}^2$ is the squared local velocity of sound in point (p_1, V_1). With $v_1 > 0$ we have

$$\boxed{v_1 > c_1.}$$
(2.78)

In point (p_2, V_2) the slope of the Rayleigh line is smaller than the tangent slope which gives the result:

$$\boxed{v_2 < c_2.}$$

(2.79)

Finally, the possible jump conditions for density, pressure and temperature at a shock wave can be obtained using the Rankine-Hugoniot equations and using the ideal gas relation $e = (\gamma - 1)RT/\mu$. Introducing the abbreviations

$$M_{a1/2} = \frac{v_{1/2}}{c_{1/2}}, \qquad c_{1/2}^2 = \left(\gamma \frac{p}{\rho}\right)_{1/2}$$

(2.80)

and algebraically manipulating (2.61), (2.67) and (2.68), we obtain the following relations:

$$\frac{\rho_2}{\rho_1} = \frac{(\gamma+1)M_1{}^2}{(\gamma+1) + (\gamma-1)(M_2{}^2 - 1)} = \frac{v_1}{v_2},$$

(2.81)

$$\frac{p_2}{p_1} = \frac{(\gamma+1) + 2\gamma(M_1{}^2 - 1)}{\gamma + 1},$$

(2.82)

$$\frac{T_2}{T_1} = \left[(\gamma+1) + 2\gamma(M_1{}^2 - 1)\right] \frac{(\gamma+1) + (\gamma-1)(M_1{}^2 - 1)}{(\gamma+1)^2 M_1{}^2}$$

$$= \frac{p_2}{p_1} \frac{\rho_1}{\rho_2},$$

(2.83)

Hence, in summary, we have the following findings:

- Since $p_2 \geq p_1$, $\rho_2 \geq \rho_1$ and $T_2 \geq T_1$, shocks are compressive.

- For the case $M_1 = 1$ there is no shock.

- For very strong shocks, $M_1 \to \infty$ and e.g. for $\gamma = \frac{5}{3}$ we have:

$$\frac{\rho_2}{\rho_1} \to \frac{\gamma+1}{\gamma-1} = 4,$$

(2.84)

i.e. the jump in density approaches a finite limiting value.

- Pressure p and temperature T are *not* limited in their maximum value in shock waves.

- Since $v_2 > v_1$, at a shock wave, the flow changes from supersonic to subsonic speed.

- There are only compression shocks, no expansion shocks, because the entropy has to increase.

2.6. General literature on shock waves

In the three-volume series *Handbook of Shock Waves* edited by Ben-Dor et. al. many different forms of shock waves and media in which shock waves appear are covered. This book also provides a historical account of the milestones of shock wave research beginning in 17th century and ending with the end of Second World War [18, 19, 20]. The classic monograph by Zel'dovich [187] gives a rather complete account of the theory and applications of shock waves. The book "Shock Compression of Condensed Matter" by Forbes [80] provides an introduction into core concepts of the shock wave physics of condensed matter, taking a continuum mechanics approach to examine liquids and isotropic solids. The review "A Review of Computational Methods in Materials Science: Examples from Shock-Wave and Polymer Physics" by Steinhauser et al. [217] exemplifies several important simulation methods and typical shock wave applications in Soft and Hard Matter systems. The chronological volume *History of Shock Waves, Explosions and Impact* is really only a history devoted mostly to empirical shock wave research that ends in 1945 [129]. In fact, until the late 1960's most of shock wave research was experimental – a consequence of the fact that the equations of motion governing shock wave generated flows are non-linear. Most computational developments in shock wave research started roughly in the 1970's when progress in computer hard- and software allowed increasingly using numerical simulations as a complementary tool to explore material behavior under shock and impact loading. An internationally recognized conference that is devoted explicitly to all phenomena pertaining to shock waves is the biannual "Topical APS Conference on Shock Compression of Condensed Matter" taking place in varying venues across the USA.

3. Multiscale modeling and simulation

Some of the most fascinating problems in all fields of science involve multiple temporal or spatial scales. Many processes occurring at a certain scale govern the behavior of the system across several (usually larger) scales. This notion of multiple hierarchies in nature, which is reflected in various approaches to multiscale modeling, can be traced back to the beginnings of modern science, see e.g. the discussions in [182, 211, 252]. In many problems of materials science the idea of multiscale modeling arises quite naturally as a *structure-property paradigm*: The basic microscopic constituents of materials are atoms, and their interactions at the microscopic level (on the scale of nanometers and femtoseconds) determine the behavior of the material at the macroscale (on the order of centimeters and milliseconds and beyond). The idea of performing material simulations across several characteristic length and time scales has therefore obvious appeal not only for fundamental research but also as a tool for technological innovation. In this chapter we will highlight several aspects of multiscale modeling in hard and soft matter before we introduce a new and conceptually straightforward coupling algorithm which combines macroscopic heat diffusion between the continuum and the atomistic domain. The motion of the atoms in the MD domain are governed by a local thermostat. This thermostat achieves the appropriate conversion between heat and MD particle velocity, generating thermal fluctuations in accordance with a Boltzmann-weighted fluctuation-dissipation theorem. This coupling method is the basis of the results of the simulations of shock wave effects in double lipid layers of coarse-grained biomembranes which we present in Chap 6.

© Springer Fachmedien Wiesbaden GmbH 2018
M. O. Steinhauser, *Multiscale Modeling and Simulation of Shock Wave-Induced Failure in Materials Science*,
https://doi.org/10.1007/978-3-658-21134-9_3

3.1. What is multiscale modeling?

With the increasing availability of very fast computers and concurrent progress in the development and understanding of efficient algorithms, numerical simulations have become prevalent in virtually any field of research [10, 44, 84, 149, 210]. Fast parallelized computer systems today allow for solving complex, non-linear many body problems directly, not involving any preceding mathematical approximations which is the normal case in analytical theory, where all but the very simplest problems of practical interest are too complex to be solved with pencil and paper. In this respect, computer simulations are not only a connecting link between analytic theory and experiment, allowing to scrutinize theories, but they can also be used as an exploratory research tool under physical conditions not feasible in real experiments in a laboratory. Computational methods have thus established a new, interdisciplinary research approach which is often referred to as "Computational Physics" or "Computational Materials Science". This approach brings together elements from diverse fields of study such as physics, mathematics, chemistry, biology, engineering and even medicine and has the potential to handle multiscale and multi-disciplinary simulations in situations relevant for real-world applications.

For example, simulations in material physics and chemistry are often focused on the investigation of lattice and defect dynamics at the atomic scale using MD and Monte Carlo (MC) methods, using force-fields (physical potentials) that are derived from solving the non-relativistic Schrödinger equation for a very limited number of atoms [38, 77, 123]. Figure 3.1 exhibits several common methods employed for atomic scale simulations along with their estimated maximum system size and the typical time scales that can be treated. The highest precision and transferability of methods is achieved with self-consistent first principles – so-called *ab-initio* – calculations.

Self-consistent Field Theory (SCF) is fundamentally based on the Hartree-Fock (HF) method which itself is a mean-field approximation (MFA); due to MFA, the energies in HF calculations are always larger than the exact energy values. The second approximation in HF

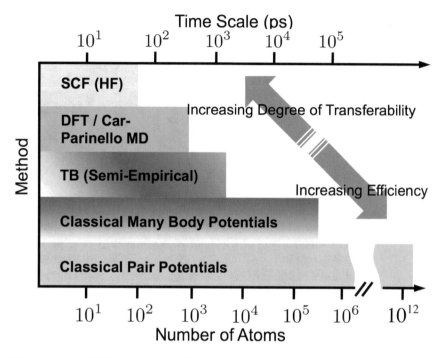

Figure 3.1.: Different atomistic methods used in physics, chemistry and materials science with their corresponding level of transferability, and a rough estimate of the number of atoms that can be simulated within a couple of days on present day supercomputing systems. Adapted from [211].

calculations is that the wave function has to be expressed in functional form; such functionals are only known exactly for very few one-electron systems and therefore, usually some approximate functions are used instead. The approximative basis wave functions which are used are either plain waves, Slater type orbitals (STO) $\sim \exp(-ax)$, or Gaussian type orbitals (GTO) $\sim \exp(-ax^2)$. Correlations are treated with Møller-Plesett perturbation theory (MPn), where n is the order of correction, Configuration Interaction (CI), Coupled Cluster theory (CP), or other methods. As a whole, these methods are referred to as *correlated calculations*.

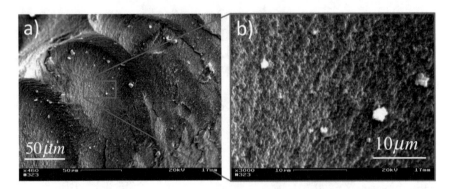

Figure 3.2.: Structural hierarchies in a micrograph section of a Al_2O_3
surface. a) Scanning electron microscopy photograph of the
fracture surface of Al_2O_3 after edge-on impact experiment
with striking speed of $v \approx 400m/s$. b) Microstructural
details of the Al_2O_3 surface. © Martin O. Steinhauser.

Density Functional Theory (DFT) is an alternative ab-initio method
to SCF. In DFT, the total energy of a system is not derived from a
wave function but rather in terms of an approximative Hamiltonian
and thus an approximative total electron density. This method uses
GTO potentials or plane waves as basis sets and correlations are
treated with Local Density Approximations (LDA).

So-called *semi-empirical methods* such as Tight Binding (TB) (e.g.
Porezag et al. [68] or Pettifort [227]) approximate the Hamiltonian
$\mathcal{H}$ used in HF calculations by approximating or neglecting several
terms (called Slater-Koster approximation), but re-parameterizing
other parts of $\mathcal{H}$ in a way so as to yield the best possible agreement
with experiments or ab initio simulations. In the simplest TB version,
the Coulomb repulsion between electrons is neglected. Thus, in this
approximation, there exists no correlation problem, but there is also
no self-consistent procedure.

Classical multi-body potentials of Tersoff or Stillinger-Weber type,
and two-body potentials, e.g. in the *Embedded Atom Method* (EAM)
or generic *Lennard-Jones* (LJ) potentials allow for a very efficient force

field calculation, but hardly can be used for systems other than the ones for which the potentials were introduced, e.g. systems with other types of bondings or atoms. Classical methods are however very important to overcome the complexities of calculations for some materials and are often used for calculating phase transitions, diffusion, growth phenomena or surface processes. In contrast to the natural sciences, materials-related simulations in the field of mechanical engineering typically focus on large–scale problems, where the N-body problem is often homogenized with approximative continuum methods, such as Smoothed Particle Hydrodynamics (SPH), the Finite Volume (FV) or the Finite Element Method (FEM) and utilizing additional, empirical constitutive relations [25, 47].

With powerful computational tools at hand, even simulations of practical interest in engineering sciences for product design and testing have become feasible. Material systems of industrial interest are highly heterogeneous and are characterized by a variety of defects, interfaces, and other microstructural features. As an example, Fig. 3.2 displays a Scanning Electron Microscopy (SEM) micrograph of the fracture surface of Aluminum Oxide (Al_2O_3) after planar impact load. Inorganic crystalline materials have structural features such as grain boundaries between crystals which are mm to μm in size, while on the atomic scale dislocations and defects such as vacancies occur. Hence, these structures have to be studied from a hierarchical perspective.

Small grain size also implies short diffusion distances, so that processes which depend on diffusion, such as sintering, are facilitated and can occur at lower temperatures than would otherwise be possible. Predicting the properties and performance of such materials under load is central for modern materials research and for product design in industry. However, due to the complexity of structural hierarchies in condensed matter on different scales, there is no single computational model or physical theory which can predict and explain all material behavior in one unified and all-embracing approach. Hence, the explicit micro structure of different important classes of materials such as metals, ceramics, or materials pertaining to soft matter (glasses or polymers) has to be incorporated in different models with delimited

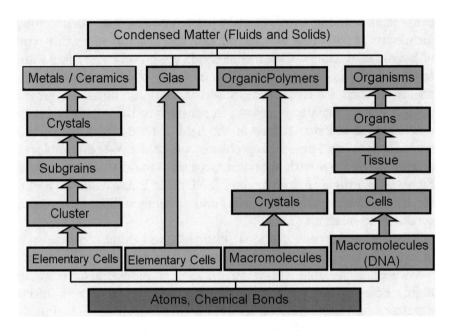

Figure 3.3.: Schematic hierarchical view of structural properties of important classes of materials, contrasting typical structural features of inorganic crystalline materials (engineering materials) and the structural features of self-organizing organic biological materials. At the nanoscale, the basic constituents condensed matter are the atoms bound together in chemical bonds. Figure adapted from [211].

validity, cf. Fig. 3.3. In many practical cases, the basic question which model shall be used for answering a specific question is the main problem.

The next step for rendering the model accessible to an algorithmic description, is the discretization of the time variable (for dynamic problems) and of the spatial domain in which the constitutive equations of the problem are to be solved. Then appropriate algorithms for solving the equations of the mathematical model have to be chosen and implemented. Before trusting the output of a newly written computer program and before applying it to new problems, the code should always be tested to the effect whether it reproduces known analytic or experimental results. This is a necessity as the correctness and plausibility of the outcome of an algorithm (usually dimensionless numbers) cannot be predicted by simply looking at the source code. The success of a computer experiment in essence depends on the creation of a model which is sufficiently detailed such that the crucial physical effects are reproduced and yet is sufficiently simple (of small complexity) for the simulation still to be feasible.

3.2. Hierarchical length and time scales

The span of length scales commonly pertaining to materials science comprises roughly 10 to 12 orders of magnitude and classical physics is sufficient to describe most of the occurring phenomena, cf. Figure 3.4. Yet, classical MD or MC methods are only valid down to length scales comparable to the typical size of atoms ($\approx 10^{-10}$m) and typically treat atoms as point particles or spheres with eigenvolume. In principle, the non-relativistic, time-dependent Schrödinger equation describes the properties of molecular (and sub-molecular) systems with high accuracy, but anything more complex than the equilibrium state of a few atoms cannot be handled at this ab-initio level. Quantum theory as a model for describing materials behavior is currently believed to be valid as a paradigm for material description on *all* length scales including the macroscopic scale, but the application of the Schrödinger equation to many particle systems of macroscopic size is not practicable due to the non-tractable complexity of the involved calculations. Hence, approximations are necessary; the larger the complexity of a system and the longer the involved time span of the investigated processes are, the more severe the required approximations are. For example, at some point, the ab-initio approach has to be abandoned completely and replaced by empirical parameterizations of the used model. Therefore, depending on the kind of question that one asks and depending on the desired accuracy with which specific structural features of the

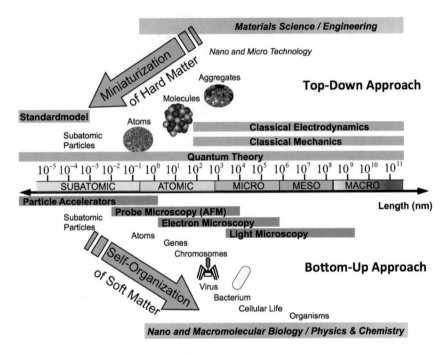

Figure 3.4.: Schematic comparing the relevant length scales in materials science according to [211]. In the field of micro- and nanotechnology one usually tries to approach the molecular level from the larger length scales, miniaturizing technical devices top-down approach), whereas nature itself always seems to follow a bottom-up approach, assembling and self-organizing its complex (soft) structures from the atomic scale to complex cellular organisms. The typical scopes of important experimental methods using microscopes are displayed as well. The validity of classical physics is limited to length scales down to approximately the size of atoms which, in classical numerical schemes, are often treated as point particles or spheres with a certain eigenvolume.

considered system are resolved, one has the choice between many different models which often can be usefully employed on a whole span of length and time scales.

Unfortunately, there is no simple "hierarchy" that is connected with a natural length scale according to which the great diversity of simulation methods could be sorted out. For example, Monte Carlo lattice methods can be applied at the femtoscale of Quantumchromodynamics (10^{-15}m) [36], at the Ångstrømscale (10^{-10}m) of solid state crystal lattices [73], or at the micrometerscale (10^{-6}m), simulating grain growth processes of polycrystal solid states [105].

The typical hierarchical structural features of materials have to be taken into account when developing mathematical and numerical models which describe their behavior. With this respect, usually one of two possible strategies is pursued: In a "sequential modeling approach" one attempts to piece together a hierarchy of computational approaches in which large-scale models use the coarse-grained representations with information obtained from more detailed, smaller-scale models ("bottom-up" vs. "top-down" approach), see Fig. 3.4 and also compare Fig. 5.1.

This sequential modeling technique has proven effective in systems in which the different scales are *weakly coupled*. The vast majority of multiscale simulations that are based on a sequential approach. Examples of this approach are abundant in literature, and it is not my intention in this work to comprehensively review the many publications in this field. The sequential approach includes practically all MD simulations whose underlying potentials are derived from ab-initio calculations, including classical coarse-grained simulations of macromolecules [119, 170, 186]. Due to the fractal nature of polymers, they exhibit self-similar structures on various length scales. This leads to unique and universal scaling properties which can be used to check the validity and quality of research codes (Chap. 5). This "Scale-Hoppingïn computer simulations of polymers is typical for coarse-grained simulations of polymer systems.

The second strategy pursued in multiscale simulations is the "concurrent" or "parallel approach". Here, one attempts to link methods appropriate at each scale together in a combined model, where the different scales of the system are considered concurrently and often communicate with some type of hand-shaking procedure [1, 2, 3]. This approach is necessary for systems, whose behavior at each scale inherently depends strongly on what happens at the other scales, for example dislocations, grain boundary structure, or dynamic crack propagation in polycrystalline materials.

3.3. Computer simulations as a research tool

Computer simulation is adding a new dimension to scientific investigation and has been established as an investigative research tool which is as important as the traditional approaches of experiment and theory. The experimentalist is concerned with obtaining factual information concerning physical states and dynamic processes. The theorist, challenged by the need to explain measured physical phenomena, invents idealized models which are subsequently translated into a mathematical formulation. As is common in theory, most mathematical analyses of the basic laws of nature as we know them, are too complex to be done in full generality and thus one is compelled to make certain model simplifications in order to make predictions. Hence, a comparison between a theoretical prediction and an experimental interpretation is frequently questioned because of the simplifying approximations with which the theoretical solution was obtained or because of the uncertainty of the experimental interpretation. For example, even the relatively "simple"laws of Newtonian mechanics become analytically unsolvable, as soon as there are more than two interacting bodies involved [193]. Most of materials science however deals with *many* ($N \sim 10^{23}$) particles, atoms, molecules or abstract constituents of a system.

Computer simulations, or *computer experiments*, are much less impaired by many degrees of freedom, lack of symmetries, or nonlinearity of equations than analytical approaches. As a result, computer simulations establish their greatest value for those systems where the gap between theoretical prediction and laboratory measurements is large.

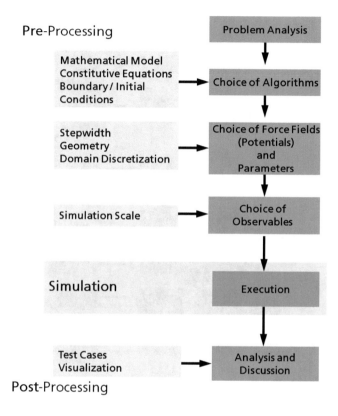

Figure 3.5.: Basic design of a computer simulation program in science and technology.

The principal design of practically all computer simulation programs for scientific purposes is displayed in Fig. 3.5: Usually, during a *pre-processing phase* some administrative tasks are done (system setup, defining initial system structure, reading in simulation parameters, initializing internal variables, etc.) before the actual simulation run is started. Analyzing data "on-the-fly" during the *simulation phase* is usually too expensive; therefore, data snapshots of the system are stored during certain preset time intervals which can later be analyzed

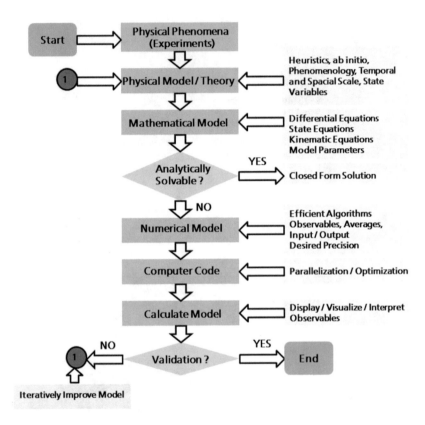

Figure 3.6.: Physical, mathematical and numerical modeling scheme
illustrated as flow chart.

and visualized during the *post-processing phase.* Very often, the
pre- and post-processing code is separated from the main simulation
code and since the mid 1990s, Graphical User Interfaces (GUIs) are
commonly used for these tasks. In UNIX environments, TCL/TK is
a classical script language used to program GUIs which is outdated
by now. Since the late 1990s, a C++ based graphical library – Qt –
is available for Open-Source developments under the GNU General
Public Licence and which is used in many modern implementations of
scientific computer codes.

The starting point for a computer simulation is the invention of an idealized adequate model of the considered physical process. This model is written in the language of mathematics and determined by physical laws, state variables, initial and boundary conditions. The question as to when a model is "adequate" to a physical problem is not easy to answer. There are sometimes many concurrent modeling strategies and it is a difficult question which aspects are essential and which ones are actually unimportant or peripheral.

This principal procedure in physical and numerical modeling is illustrated schematically in the flowchart of Fig. 3.6: Starting from the experimental evidence one constructs physical theories for which a mathematical formulation usually leads to differential equations, integral equations, or master (rate) equations for the dynamic (i.e. time dependent) development of certain state variables within the system's abstract state space. Analytic solutions of these equations are very rarely possible, except when introducing simplifications usually involving symmetries. Thus, efficient algorithms for the treated problem have to be found and implemented as a computer program. Execution of the code yields approximate numerical solutions to the mathematical model which describes the dynamics of the physical "real" system. Comparison of the obtained numerical results with experimental data allows for a validation of the used model and subsequent iterative improvement of the model and of theory.

3.4. Simulation methods for different length and time scales

The first scientific simulation methods ever developed and implemented on working electronic computers were MC and MD methods, fully rooted in classical physics [7, 8, 9, 22, 160, 161, 162, 185]. Many problems of classical MD techniques lie in the restriction to small (atomistic and microscopic) length and time scales. In atomistic MD simulations of hard matter, i.e. crystalline systems which are mostly governed by their available energy states, the upper limit on todays hardware is typically a cube with an edge length of a few hundred nanometers simulated for a few nanoseconds. However, on the largest available computer systems, MD simulation techniques have passed the 1 trillion (10^{12}) particle boundary and corresponding shock wave simulations in crystal lattices have been performed [85]. However, such large simulations are really only possible on the largest super computer systems in the world, are heavily hardware-optimized and not readily transferable to different hardwares with the same parallel scaling.

With *coarse-grained models*, where the individual MD particles represent complete clusters of atoms, molecules or other constituents of the system, this limit can be extended to microseconds or even seconds. In this respect, soft matter systems such as polymers, which are very long macromolecules, constitute a very interesting class of materials due to their intrinsic universal scaling features [229, 206] which are a consequence of their fractal properties [55]. Macroscopic physical properties of materials can be distinguished in:

- *static equilibrium properties*, e.g. the radial distribution function of a liquid, the potential energy of a system averaged over many time steps, the static structure function of a complex molecule, or the binding energy of an enzyme attached to a biological lipid membrane.

- *dynamic or non-equilibrium properties*, such as diffusion processes in biomembranes, the viscosity of a liquid, or the dynamics of the propagation of cracks and defects in crystalline materials.

Many different properties of materials are determined by structural hierarchies and processes on multiscales. An efficient modeling of the system under investigation therefore requires special simulation techniques which are adopted to the respective problems. Table 3.1 provides a general overview of different simulation techniques used on various length scales in materials science along with some typical applications. This division is primarily based on a spatial, rather than a physical classification.

Table 3.1.: Customary classification of length scales. Displayed are also typical scopes of different simulation methods and some typical applications pertaining to the respective scale. Table adapted from [217].

Typical Simulation Methods	**T**ypical Applications
Mesoscopic/Macroscopic	
Hydrodynamics [94]	macroscopic flow
Finite Element Methods [17, 21]	plasticity
SPH [149, 166]	fracture mechanics
Finite Difference Methods [5, 45]	aging of materials
Cluster & Percolation Models	fatigue and wear
Microscopic/Mesoscopic	
MD and MC using	complex fluids
effective force fields [210, 222]	soft matter
Dissipative Particle Dynamics [70, 71, 106]	granular matter
Cellular Automata [245]	grain growth
Lattice-Boltzmann [134]	interface motion
Dislocation Dynamics [37, 56, 251]	dislocations
Discrete Element Method [50]	grain boundaries
Atomistic/Microscopic	
Molecular Dynamics [7, 8, 9]	equations of state
Classical Monte Carlo [161, 162]	Ising model, DNA
Hybrid MD/MC [23, 24]	polymers, rheology
Embedded Atom Method [53, 54, 79]	transport properties
Electronic/Atomistic	
Self-Consistent Hartree-Fock [78, 100]	crystal ground states
Self-Consistent DFT [102, 123, 124]	NMR, IR spectra
Car-Parinello (ab initio) MD [38]	molecular geometry
Tight-Binding [68]	electronic properties
Quantum Monte Carlo (QMC) [16]	chemical reactions

3.5. Computer programs and implementation details

The computer simulation results presented in this work have been generated with a number of different simulation codes and different simulation techniques based on discrete elements, e.g. particles, and on the approximation of a of a system as a continuum with infinite degrees of freedom. It is not my intention in this work to review the basic methods of various simulation techniques. In parts, this has been already done elsewhere [210, 211, 217]. In principle, for producing the results presented in this work, the following simulation techniques were used:

- Finite Element Method (FEM) in Chap. 4,

- Smoothed Particle Hydrodynamics (SPH) in Chaps. 3, 5 and 6,

- Molecular Dynamics (MD) in Chaps. 3, 4, 5 and 6,

- Discrete Element Method (DEM) in Chap. 4,

- Monte Carlo (MC) in Chap. 4,

- Dissipative Particle Dynamics (DPD) in Chaps. 3, 5 and 6.

The MD, DEM, SPH, MC, and DPD methods are implemented in the research code MD-CUBE, a simulation software suite authored by M. O. Steinhauser [206]. The tool MD-CUBE was originally developed by M. O. Steinhauser in a national Fraunhofer research program "MMM-Tools: Tools for Consistent Multiscale Modeling of Materials" during the years 2003-2005 with a budget of 3.5 million Euros that was directed by the author of this work as principal investigator and managing director. During the last decade the program has been extended to a full software suite with several modules that implement various simulation methods that can be linked to the program either as library or via compiler options, see Fig. 3.7. The Monte Carlo technique for optimizing our generated power diagrams of 3D microstructures

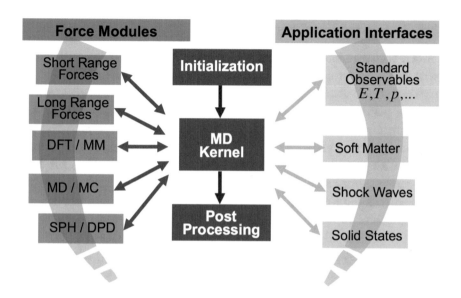

Figure 3.7.: Basic architecture of the software suite MD-CUBE.

shown in Chap. 4, was implemented in the research code POLYGRAIN
which is part of the MD-CUBE software suite. The impact simulations
based on the power diagrams in Chap. 4 were produced with standard
commercial FEM software LS-DYNA.

3.5.1. Reduced simulation units

Unless otherwise stated, we employ the standard system of reduced
units [10], defined by the unit of energy ϵ and the unit of length σ. As
example of these units we note the reduced temperature $T^* = k_B T/\epsilon$,
reduced number density $\rho^* = \rho\sigma^3$, reduced pressure $p^* = p\sigma^3/\epsilon$, and
reduced simulation time $t^* = t\sqrt{\epsilon/m\sigma^2}$. For simplicity in our notation,
we drop the superscript * when writing down quantities in reduced
simulation units.

3.5.2. Shock wave generation

Shock waves in the simulations presented in this work were initiated with a momentum reflecting mirror [104], a standard method for generating shock waves in simulation studies of hard condensed matter. Here, a piston of infinite mass moves with constant velocity v_p against the target material. All particles coming into contact with the piston surface are reflected. This method is quite similar to standard shock wave experiments, where a static target material is hit by a fast-moving impactor [215]. Upon impact, the target material is compressed and the resulting steep density gradient initiates a shock wave. This method allows for producing fast shock waves up to $10^4 \mathrm{ms}^{-1}$ with a well defined shock wave front and good numerical stability. An alternative method was used by Koshiyama et al. [125], who added a constant hypersonic velocity to all particles enclosed within a certain region of the simulation volume. The latter method is computationally less efficient compared to the momentum reflecting mirror as a comparatively larger number of particles is required for initiating the shock wave.

3.6. Coupling the atomic and continuum domain

In this section we present a new thermodynamically consistent multiscale coupling scheme that combines an atomic description of matter based on the MD method, with a continuum description based on the SPH method.

The multiscale coupling of methods which operate on different time- and length scales is a rather active topic of current research allowing physicists, chemists and material scientists to gain insight into phenomena, where microscopic processes determine macroscopic effects. One area of application includes crack nucleation and propagation in materials, leading to material fatigue and, eventually, failure [4, 218, 220].

Polymers and biological macromolecules form another interesting research area for the application of multiscale methods due to their fractal nature. Polymers exhibit self-similar structures on different length- and time scales, which – along with coarse-graining (Chap. 5) – leads to universal scaling properties of these systems that can be seen already in relatively short linear polymer chains with $N \approx 50$ monomers.

Over the past three decades, numerous different atomistic-continuum coupling strategies have been devised [3, 33, 66, 119, 131, 154, 168, 169, 171, 247], which I do not intend to comprehensively review here. All of these methods offer advantages in specific situations, but lack the necessary generality to be applied as a standard tool. In particular, the problem of describing heat exchange across the continuum-atomistic interface has not yet been solved to a satisfying degree. The existing body of literature considering coupling of atomistic and continuum scales has almost exclusively considered isothermal processes, for which heat exchange can be neglected. However, a huge range of interesting problems are of transient, discontinuous and therefore non-equilibrium nature: shock induced deformation, or unsteady shear flow are examples. These processes involve, by definition, steep gradients of temperature and density which need to be accurately described by a scale-bridging simulation method in order to allow for faithful modeling.

The conceptual problem with formulating heat flux between continuum and atomistic domains is rooted in the fundamentally different representation of these domains: In the continuum approach, one discretizes a continuous field of state variables at discrete spatial locations (integration nodes), between which the fluxes of heat and mass are numerically evaluated according to, e.g. the set of Navier-Stokes equations (2.15). They are closed with a constitutive equation of state which links density and pressure, along with a description of transport properties such as viscosity. The atomistic approach taken in MD is to use classical Newtonian particles which carry mass and interact with each other via distance- and orientation-dependent forces. In MD, momentum exchange is achieved through the many-body interactions between particles, which determines the time evolution of position and velocity. Because no other degrees of freedom except those associated with the particle kinetic energy are present, temperature depends solely on momentum, as expressed by the equipartition theorem:

$$\langle T_{\text{kin}} \rangle = \frac{1}{f k_B} \sum_{i=1}^{N} m_i \vec{v}_i^{\,2}. \tag{3.1}$$

Here, $\vec{v}_i$ and m_i are velocity and mass of particle i, while f and k_B are the system's number of degrees of freedom (we consider only translation here), and Boltzmann's constant, respectively. The number N denotes the ensemble size, which has to be taken large enough in order to be meaningful in a macroscopic interpretation. We have explicitly marked this temperature expression with the index "kin" to emphasize that the appropriate temperature definition in MD is a kinetic temperature. On the other hand, the continuum expression for temperature is based on the concept of internal energy e and heat capacity $c_V(e_i)$, both defined individually for the corresponding integration node i:

$$T_i = \frac{e_i}{c_V(e_i)}. \tag{3.2}$$

Thus, temperature in an atomistic description relates to average particle momentum, while it is just a state variable in the continuum representation, with no relation to the momentum of the corresponding continuum integration node. The challenge for a scale-bridging simulation technique is to effect both momentum exchange and heat flux across the continuum-atomistic interface. Momentum exchange can be incorporated relatively easy by choosing a spatial node discretization density in accordance with the number density of the MD particles, and using MD forces between continuum nodes and MD particles. However, in an adiabatic situation where the MD particle's velocities are not coupled to an external thermostat, this approach leads to reduction of the temperature in the MD domain, and eventually freezing, at the cost of an increased temperature in the continuum domain. This behavior is due to the momentum exchange, which accelerates the integration nodes. These nodes are subsequently slowed down by the inherent inner friction (viscosity) operating in the continuum domain, unidirectionally converting the nodes' kinetic energy into internal energy. What is lacking on the continuum side are the thermal fluctuations, which convert back and forth between internal energy and node velocities in accordance with the appropriate Boltzmann distribution. Various approaches have been published to incorporate these fluctuations into a continuum description, with the work of Español [70] being the currently most stringent and promising.

In Chap. 6 we use the method presented in Sec. 3.6.1 to perform multiscale simulations of the effects of shock wave destruction in a phospholipid bilayer surrounded by a fluid (water). Here, in the following we present equilibrium results of our new multiscale coupling scheme for a shock tube simulation where the tube is filled with MD and SPH particles. The validity of our new coupling scheme is demonstrated in Sec. 3.7 by showing that salient observables, such as the Maxwell-Boltzmann distribution of momenta, and the correct propagation of a shock wave across continuum and MD domains, free of numerical artifacts, are fulfilled.

3.6.1. Dissipative particle dynamics at constant energy

A correct continuum/atomistic coupling algorithm needs to describe heat flux between the two domains such that the continuum variable internal energy e_i is locally linked to the atomic particle velocity. Such a thermostat is given by the formulation of Dissipative Particle Dynamics at constant Energy (DPDE) due to Avalos and Mackie [13], and Español [70], which is applied to the MD particles. DPDE is an extension to Dissipative Particle Dynamics (DPD) [71, 106], which is a coarse-graining technique for MD. DPD mimics the complex particle dynamics of a fully detailed system by incorporating stochastic fluctuations in the equations of motion of the corresponding coarse-grained system which uses much less particles. In DPD, atomistic representations are reduced to models with only a small number of particles, and the „fast" variables related to the coarse-grained (integrated out)

degrees of freedom are replaced by random and dissipative forces,
which mimic thermal fluctuations. Because these interactions only
control temperature and dynamics, conservative forces $\vec{F}_{ij}{}^C$, derived
from the interaction potentials need to be included. The random and
dissipative interactions between particles i and j read [231]:

$$\vec{F}_{ij}{}^R = \zeta W(r_{ij})\vec{e}_{ij}\frac{\xi_{ij}}{\sqrt{\Delta t}}, \tag{3.3}$$

$$\vec{F}_{ij}{}^D = -\gamma W^2(r_{ij})(\vec{v}_{ij} \cdot \vec{e}_{ij})\vec{e}_{ij}, \tag{3.4}$$

where $\vec{r}_{ij} = \vec{r}_i - \vec{r}_j$, $r_{ij} = |\vec{r}_{ij}|$, $\vec{e}_{ij} = \vec{r}_{ij}/r_{ij}$, and $\vec{v}_{ij} = \vec{v}_i - \vec{v}_j$,
with $\vec{r}$ and $\vec{v}$ being the particle's position and velocity, respectively.
The ξ_{ij} are symmetric random variables with zero mean and unit
variance, which are independent for different pairs of particles and
different times and Δt is the time step of the integration. $W(r_{ij})$
is a weight function with compact support, determining the DPD
interaction radius between two particles. A fluctuation-dissipation
theorem relates these functions, as well as the amplitudes ζ and γ of
the random and viscous dissipative forces, in order to preserve the
canonical phase space measure:

$$\gamma = \frac{1}{2k_B T}\zeta^2. \tag{3.5}$$

We employ the standard linear DPD weight function with interaction
range r_c^{DPD},

$$W(r_{ij}) = \begin{cases} 1 - r_{ij}/r_c^{\mathrm{DPD}} & r_{ij} \leq r_c^{\mathrm{DPD}}, \\ 0 & r_{ij} > r_c^{\mathrm{DPD}}. \end{cases} \tag{3.6}$$

ζ is a free parameter which controls the amount of momentum exchange
between pairs of particles per time step and has direct influence on
the dynamics of the system. With the above definitions, the DPD
equations of motion read:

$$d\vec{r}_i = \vec{v}_i dt \tag{3.7}$$

$$d\vec{v}_i = \frac{1}{m_i}\left[\sum_{j \neq i}\left(\vec{F}_{ij}{}^C + \vec{F}_{ij}{}^D + \vec{F}_{ij}{}^R\right)\right] dt. \tag{3.8}$$

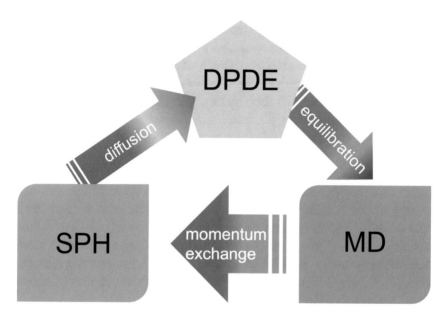

Figure 3.8.: Sketch of the proposed continuum–atomistic coupling algorithm. Note that – at equilibrium – the flow is directed as shown because continuum viscosity irreversibly transforms nodal velocity into internal energy.

Up to this point, the temperature is fixed in the fluctuation-dissipation theorem, with the effect that DPD is actually a thermostat, which adds to, and removes energy from the system, until the particles' kinetic energy is in agreement with the specified temperature. Thus, DPD is strictly an *isothermal* method. In contrast, DPDE associates a *local internal temperature* T_i with each particle i, given by an internal energy variable e_i and a heat capacity c_v, such that $T_i = e_i/c_V(e_i)$. DPDE is therefore a *mesoscopic* simulation method, incorporating both a continuum description of temperature (3.2) and per-particle degrees of freedom like MD. With only a *local* temperature dependence, DPDE is able to describe temperature gradients as needed for the simulation of non-equilibrium phenomena, such as shock wave propagation in a medium.

We thus propose a coupling strategy as follows [83]: In the continuum domain, the time evolution of the system is given by constitutive equations, and the domain is spatially discretized into points, which serve as integrations nodes for the partial differential equations of continuum mechanics. To this domain, a region with atomistic length-scales and corresponding particle dynamics is coupled. This region is described by classical MD, i.e., interactions between pairs of particles following microscopic classical laws are used to describe physical properties. Because the concept of temperature is fundamentally different in both domains – average particle velocity in MD vs. internal energy in the continuum – the local DPDE thermostat is used in the MD domain to achieve dynamic equilibrium between both temperature definitions. In addition, macroscopic heat conduction is used to model heat diffusion between MD/DPDE particles and continuum integration nodes. Figure 3.8 provides a schematic overview of our proposed continuum–atomistic coupling scheme.

The DPDE pair interaction, which acts between pairs of MD particles, is given by the thermal force

$$\vec{F}_{ij}^{\text{DPDE}} = \left(\sqrt{2 k_B T_{ij} \gamma_{ij}} W_{ij} \frac{\xi_{ij}}{\sqrt{\delta \tau}} - \gamma_{ij} W_{ij}^{2} \vec{v}_{ij} \hat{r}_{ij} \right) \hat{r}_{ij}, \qquad (3.9)$$

Here, $T_{ij} = 2(1/T_i + 1/T_j)^{-1}$ is the average local temperature of the interacting particles. The effect of the thermal force is accelerating or slowing down pairs of particles such that the particle velocity is in agreement with the local temperature, (3.1) and (3.2). The time-evolution of the particles' internal energy follows from imposing conservation of total energy, but needs to be evaluated using Itô calculus [70] due to the stochastic nature of the force in (3.9):

$$\begin{aligned}
\frac{\delta e_i}{\delta \tau} &= \frac{1}{2} \sum_j \left\{ \gamma_{ij} W_{ij}^2 \left[\left(\vec{v}_{ij} \hat{r}_{ij} \right)^2 - k_b T \left(\frac{1}{m_i} + \frac{1}{m_j} \right) \right] \right. \\
&\quad \left. - \sqrt{2 k_B T_{ij} \gamma_{ij}} W_{ij} \vec{v}_{ij} \hat{r}_{ij} \frac{\xi_{ij}}{\sqrt{\mathrm{dt}}} \right\}. \qquad (3.10)
\end{aligned}$$

We note here that the original derivations of DPDE also include macroscopic thermal conduction. However, following the arguments presented in [225], we neglect these terms here as the evolution of the internal energies is expected to be dominated by dissipative forces at the shock front. Additionally, as the individual mesoscopic particles considered in our simulations only represent approximately 10 to 50 atoms, we expect macroscopic heat diffusion processes to be realistically modeled by momentum exchange only.

3.6.2. SPH approximation of the continuum

We restrict ourselves here to SPH for describing the continuum domain, although the proposed coupling algorithm is also suitable to other continuum discretization methods. Coupling of SPH and MD is most appealing from a conceptual perspective, as both are mesh-free methods, with their time-evolution governed by Newton's equations of motion. In SPH, a continuum domain is discretized by employing a number of Lagrangian integration nodes carrying mass m_i and internal energy e_i. Together with the heat capacity c_V, a local temperature $T_i = e_i/c_V(e_i)$ is defined in a similar way as in the method used in the DPDE thermostat. The mass density at each integration node (SPH "particle") is calculated from the equation

$$\rho_i = \sum_j m_j W_{ij}, \qquad (3.11)$$

where the sum extends over all integration nodes within the range of the weighting function W_{ij}, centered at integration node i. With temperature and mass density defined, a yet to be defined equation of state provides the pressure $p_i \mapsto p(\rho_i, T_i)$ from which the forces between integration nodes follow [107]:

$$\vec{F}_{ij} = -m_i m_j \left(\frac{p_i}{\rho_i^2} + \frac{p_j}{\rho_j^2} + \Pi_{ij} \right) \vec{\nabla} W_{ij}. \qquad (3.12)$$

In the above equation, $\vec{\nabla} W_{ij}$ is the spatial derivative of the weighting function $W(|\vec{r}_{ij}|)$ with respect to $\vec{r}_{ij}$, and Π_{ij} is a viscous term, which we take to be the standard artificial viscosity due to Monaghan [165], with effective kinematic viscosity ν. The corresponding change in internal energy of particle i per time-step due to the SPH forces is [107]:

$$\frac{\delta e_i}{\delta \tau} = -\frac{1}{2} \sum_j m_i m_j \left(\frac{P_i}{\rho_i^2} + \frac{P_j}{\rho_j^2} + \Pi_{ij} \right) \vec{v}_{ij} \vec{\nabla} W_{ij}. \qquad (3.13)$$

3.6.3. Macroscopic heat flow

The continuum expression for heat conduction is

$$\vec{J} = -\kappa \vec{\nabla} T. \qquad (3.14)$$

The flux of internal energy (heat) $\vec{J}$ is proportional to the temperature gradient $\vec{\nabla} T$, with conduction coefficient κ. The solution for the time evolution of this flow is a diffusion equation, which has been cast into a SPH discretized form as:

$$\frac{\delta e_i}{\delta \tau} = -\sum_j \frac{m_i m_j}{\rho_i \rho_j} \frac{(\kappa_i + \kappa_j)(T_i - T_j)}{|\vec{r}_{ij}|} \vec{r}_{ij} \vec{\nabla} W_{ij}. \qquad (3.15)$$

The conduction coefficient κ is related to the heat diffusion coefficient $\alpha = \kappa c_V \rho / m$.

3.7. Proof of principle: SPH/MD coupling in a shock tube

In order verify our approach proposed here, we consider a Lennard-Jones (LJ) fluid defined by the familiar pair interaction

$$u(r_{ij}) = 4\epsilon \left[\left(\frac{\sigma}{r_{ij}} \right)^{12} - \left(\frac{\sigma}{r_{ij}} \right)^6 \right]. \qquad (3.16)$$

Due to the computational efficiency of this pair potential, its equation of state has been determined with great accuracy. We therefore have the fortunate situation that both continuum and MD methods yield the same results, which renders this an ideal test case for studying the coupling of these methods. On the continuum side, we use SPH particles with a LJ equation of state published in [190]. For the MD/DPDE particles, interactions are given by the pair potential (3.16) and DPDE forces are provided by (3.9).

The LJ pair potential is cut and shifted at 5σ and the DPDE amplitude is set to $\gamma = 0.5$. For the weighting function we use Lucy's choice [150]

$$W(|\vec{r}_{ij}|) = \begin{cases} \frac{105}{16\pi h^3} \left[1 + 3\frac{|\vec{r}_{ij}|}{h}\right] \left[1 - \frac{|\vec{r}_{ij}|}{h}\right]^3 & |\vec{r}_{ij}| < h \\ 0 & |\vec{r}_{ij}| \geq h, \end{cases} \quad (3.17)$$

with $h = 2.4\sigma$. Heat conduction according to Eq. (3.15) exists across all particles, with a thermal diffusivity coefficient $\alpha = 10$. The effective kinematic viscosity acting between SPH particles was set to $\nu = 0.3$, in agreement with the transport properties measured by MD simulations of the LJ potential at the state points (ρ, T) that are considered here. We note that simulation results are rather insensitive to the specific choice of α and γ, provided that the DPDE conversion rate of internal/kinetic energy is faster than the dissipation rate due to the viscosity, such as not to create a bottleneck in the scheme depicted in Fig. 3.8. The cross-interaction between SPH and MD/DPDE particles is simply a weighted sum of both MD/DPDE and SPH forces, with weighting coefficients of one half for each type of interaction. All particles are assigned a value for the reduced mass of $m = 1$. Integration of the equations of motion is performed using the velocity Verlet algorithm.

3.7.1. Shock tube: equilibrium properties

With the above defined force properties, a system with $\rho = 0.9$, consisting of $N = 230,400$ particles, is set up on a simple cubic lattice with x, y, z extending over $400 \times 24 \times 24$ lattice points, centered about the origin (0,0,0). From these, the central $N_{MD} = 86,976$ lattice

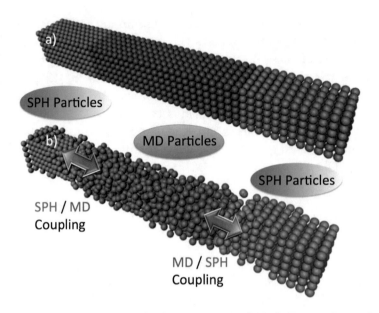

Figure 3.9.: Shock tube setup for testing the SPH/MD coupling scheme.
a) Initial setup. b) Snapshot after a couple of time steps,
showing strong fluctuations of particles in the MD/DPDE
region, whereas the SPH particles (which are just integration nodes at which the continuum conservation equations are solved) do not move. The kinetic energy of the
MD particles is transferred into internal energy of the SPH
particles of the continuum and vice versa.

points are assigned to be of MD/DPDE type, and the rest of SPH
type, see Fig. 3.9. Periodic boundary conditions were used along
all three Cartesian directions. All particles are assigned a constant
heat capacity $c_V = 20$ and internal energy $e = 12$, yielding an initial
internal temperature $T = 0.6$ and zero initial velocity. Using a time
step $\Delta t = 10^{-4}$, the system is then run for 5×10^6 time steps. This
choice of time step is required for good energy conservation within
the MD/DPDE region; the corresponding Courant-Friedrichs-Lewy
(CFL) criterion for the SPH region predicts $\Delta t \approx 0.1 h/c_0 \simeq 5 \times 10^{-2}$,

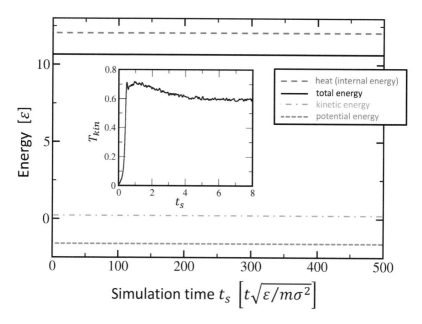

Figure 3.10.: Time evolution of potential, kinetic, and internal energies corresponding to the shock tube problem of Sec. 3.7.1. The inset shows the initial equilibration of the MD/DPDE kinetic temperature which is proportional to the kinetic energy T_{kin} [83].

given that the equilibrium speed of sound at the considered state point is $c_0 \simeq 4.7$. After a time $t \simeq 0.5$, the velocity average within the MD/DPDE particle region first attains a value corresponding to $T \simeq 0.6$, followed by a mild overshoot which is damped out after $t \simeq 4.0$.

Figure 3.10 shows the time-evolution of kinetic, potential, and internal energies as well as the kinetic temperature of the MD/DPDE particles only. We note that conservation of total energy in the DPDE algorithm is not as good as one usually expects from MD and Velocity-Verlet time integration [10] because (3.10) is only accurate to $\mathcal{O}(\Delta t)$. Nevertheless, this very long run conserves total energy with an

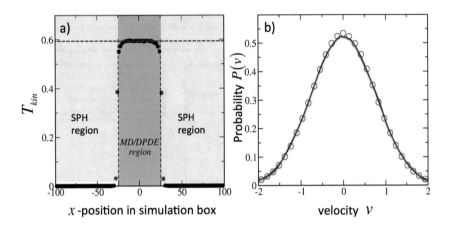

Figure 3.11.: a) Profile of the kinetic temperature $\langle T_{\mathrm{kin}} \rangle$ along the
x-axis of shock tube. b) Comparison of the Maxwell-
Boltzmann velocity distribution, (3.19), (full line) with
the measured results for the MD particles (open symbols).
Figure adapted from [83].

accuracy of 100 parts per million, an error which is hardly discernible
from Fig. 3.10.

Following equilibration, data are accumulated in a histogram of
particle velocities with bin width 2σ, resolved along the x-coordinate.
We expect these velocities to correspond directly to the temperature
definition within the MD region, according to (3.1). Neglecting any
change in potential energy (as justified by Fig. 3.10), this temperature
can be estimated a priori by redistributing the internal energy among
the internal heat capacity and the translational degrees of freedom of
the MD particles:

$$\langle T_{kin} \rangle = \frac{N \times 12\epsilon}{(N c_V + N_{MD} \frac{3}{2} k_B)} \simeq 0.593 \qquad (3.18)$$

We expect zero particle velocity within the continuum domain, due
to the irreversibility of macroscopic viscosity. Figure 3.11 a) shows
that these expectations are met with a steep change of the particle
velocities at the interface between MD and SPH regions, smoothed

over a few particle spacings. Within the center of the MD region, a kinetic temperature corresponding very closely to the estimate given above is reached. Additionally, we consider the velocity distribution of the MD/DPDE particle domain. From equilibrium thermodynamics, we expect this distribution to be given by the Maxwell-Boltzmann distribution

$$P(|\vec{v}|) = \left(\frac{1}{2\pi m k_B T}\right)^{3/2} \exp\left(-\frac{m|\vec{v}|^2}{2k_B T}\right), \tag{3.19}$$

a necessary condition for correct microscopic dynamics. As shown in Fig. 3.11 b), the simulation results accurately follow the theoretical prediction.

3.7.2. Shock tube: dynamic properties

In order to demonstrate the applicability of our proposed coupling scheme to situations far away from equilibrium, we consider generation and propagation of a shock wave in a shock tube, which is partially described by continuum SPH and partially by MD [83]. This case provides a stringent test for the applicability of the algorithm to steep gradients in pressure, density, and temperature. The initial system configuration is as described above, with the only difference that momentum-reflecting mirrors are applied at the x-axis boundaries, and that MD/DPDE particles are given initial random velocities corresponding to $\langle T_{kin}\rangle = 0.6$. Following an equilibration run of $100,000$ time steps, a shock wave is initiated by adding a constant velocity $-v_p$ to the x-component of each particle's velocity. Thus, the entire system is treated as a bar which impacts a rigid wall. This induces a shock wave as particles hit and pile up against the left wall, see Fig. 3.12. The shock front, i.e., the discontinuity at which the mass density increases suddenly, moves with speed v to the right, such that the shock front velocity is given in a co-moving frame as $v_s = v + v_p$. The shock wave first propagates through the SPH region, then through the MD/DPDE region, and again through an SPH region. To our best knowledge, we are the first researchers who applied the basic equations

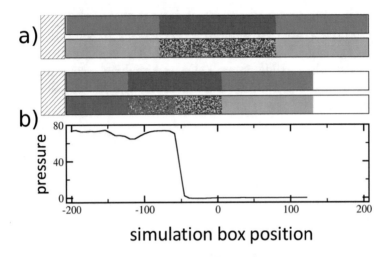

simulation box position

Figure 3.12.: Snapshots of shock wave propagation in a shock tube [83]. a) yz-projection of the simulation box; the shaded area on the left denotes a momentum-reflecting mirror. All particles move left with particle velocity of $v_p = 6$ in reduced units. The snapshot is shown in two different representations, with the upper part indicating particle type: bright (respectively red): SPH; dark (respectively blue): MD), and the lower part of the figure represents the local pressure: dark (resp. blue): low pressure, bright (resp. red): high pressure. Note that the average pressure in the MD region equals the pressure in the SPH region. b) shows the simulation at $t = 12.5$ in reduced units, after the particles have moved and compressed against the left wall, with a corresponding pressure increase. The shock wave front has reached the center of the MD particle region, as indicated by the pressure profile.

of DPDE by Pep Español [70] to a useful problem of chemical physics and solved this problem by implementing the method as a thermostat for coupling the continuum with the atomic domain.

Part II.

Hard matter

4. Shock wave failure in granular materials

Understanding the microstructural features of polycrystalline materials such as high-performance ceramics (HPCs) is a prerequisite for the design of materials with desired superior properties, such as high toughness or strength. On the length scale of a few microns to a few hundreds of microns, many materials such as glass, concrete or ceramics exhibit a polyhedral granular structure which is known to crucially influence their macroscopic mechanical properties. With ceramics, the specific shape and size of these polycrystalline grain structures is formed during a sintering process where atomic diffusion plays a dominant role. Usually, the sintering process results in a dense microstructure with grain sizes ranging from below $1 \mu m$ to several hundreds of micrometers. Using a nanosized fine-grained granulate as a green body along with an adequate process control it is possible to minimize both, the porosity ($< 0.05\%$ in volume) as well as the generated average grain size ($< 1 \mu m$). It is known that both leads to a dramatic increase in hardness which outperforms most metal alloys at considerably lower weight. It is striking that producing very small grain sizes in the making of HPCs below 100nm results again in decreasing hardness [130]. Hence, there is no simple connection between grain size and hardness of a polycrystalline material.

Today, one is still compelled to search for the optimal microstructure for a specific application by intricate and expensive experimental 'trial-and-error' studies. In order to overcome this situation by computational modeling and numerical simulations, a detailed and realistic modeling of the available experimental microstructures in three dimensions (3D) is a basic requirement. Follow-up finite element (FE)

© Springer Fachmedien Wiesbaden GmbH 2018
M. O. Steinhauser, *Multiscale Modeling and Simulation of Shock Wave-Induced Failure in Materials Science*,
https://doi.org/10.1007/978-3-658-21134-9_4

a) Experiment   **b) Power Diagrams**   **c) Particle Network**

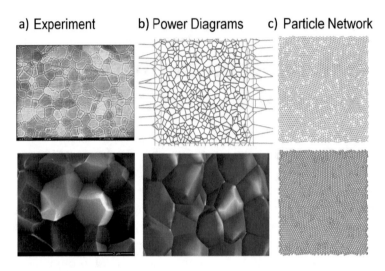

Figure 4.1.: a) Polished micrograph section of an aluminum-oxide
ceramic (Al_2O_3), which is obtained by etching the surface.
The typical average grain size of this type of ceramic is below $1.0\mu m$. The lower picture shows a SEM image of the
same specimen. b) A two-dimensional (2D), virtual section
of a three-dimensional (3D) PD, representing a polyhedral
structure as described in [133]. The figure below exhibits
3D granular structure obtained from a calculated and optimized PD. Note the rough surface structure of the PD
and the similarity to the structure of the SEM photograph
displayed in a). c) Modeling of microstructural features of
granular solids using a multiscale method based on discrete
elements (in our case: spherical particles), see [211, 215].
Top: 2D representation of the network density of randomly
overlapping particles. Bottom: Simulation snapshot of the
same network displaying the overlapping particles. Adapted
from [215]. All figures are © Martin Steinhauser.

analyses can then investigate the observed crack patterns and the specific failure of these granular materials. To achieve this goal, one has
to fit computer-generated microstructures as close as possible to the
structures observed in experiments, see Fig. 4.1. For numerical shock

wave simulations, this 3D microstructure structure is meshed using a tetrahedral volume mesh (not shown in the figure). With numerical investigations taking explicitly into account the microstructural details, one can expect to achieve a considerably enhanced understanding of the structure-property relationships of materials [41].

New ideas for the microstructure generation of granular materials in 3D are presented here which are based on two completely different approaches [89, 215, 219, 220]: One approach uses the mathematical theory of the power diagram (PD) which is an extension of the ordinary Voronoi diagram (VD) and allows for subsequent FE analyses under shock loading conditions, as discussed in Sec. 4.1. With this approach one tries to mimik as close as possible the experimental set-up of the edge-on-impact geometry which is used to generate shock wave impact experiments, which lead to fracture and ultimate failure of HPCs. We discuss these experiments in Sec. 4.2. The second modeling approach is based on a mesh-free (i.e. particle-based) multiscale model presented in Sec. 4.3, and which is implemented using the DEM.

4.1. Polyhedral cell complexes and power diagrams

The microstructure of densely sintered ceramics can be considered in very good approximation as a tessellation of $\mathbb{R}^2$ with convex polyhedra, i.e. as a polyhedral cell complex, cf. Fig. 4.1. A direct, primitive discretization of the SEM micrograph as shown in Fig. 4.1 a) into equal-spaced squares in a 2D mesh can be used for a direct simulation of material properties using the FEM, cf. Fig. 4.2. However, with this computational approach, the grain boundaries on the microscale have to be modeled explicitly with very small elements of finite thickness. Thus, the influence of the area of the interface is unrealistically overestimated in light of the known fact that grain boundaries, which constitute an area of local disorder, often exhibit only a thickness of a few layers of atoms [31]. Moreover, a photomicrograph is just *one* 2D sample of the real microstructure in 3D, hence the value of its

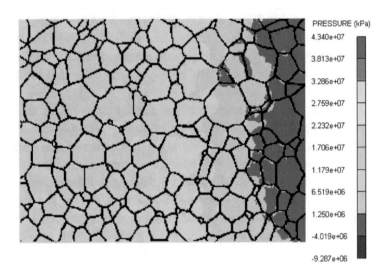

Figure 4.2.: FE simulation of a digitized 2D model of the SEM micrograph of Al_2O_3 from Fig. 4.1a. A shock wave is traveling through the material from left to right. The plane of the micrograph has been sectioned into 601×442 equal-spaced squares which are used as finite elements. The nodes of the upper and lower edge have been assigned $\vec{v} = 0$ as boundary condition, whereas the leftmost element nodes of the sample are given an initial speed of $v_x = 500m/s$. Taken from [215].

explicit rendering is very questionable. Finally, with this approach there is no 3D information available at all. While experimentally measured microstructures in 3D are generally not available for ceramic materials, some reports about measured microstructures of steel were published [137, 143, 144]. Nevertheless, these experiments are expensive and their resolution as well as the number of measured grains still seem to be poor [144].

A different way of generating microstructures, is based on classical Voronoi diagrams in d-dimensional Euclidean space $\mathbb{E}^d$ and their duals, the Delaunay triangulations, see Fig. 4.3. Both constitute important models in stochastic geometry and have been used in various scientific

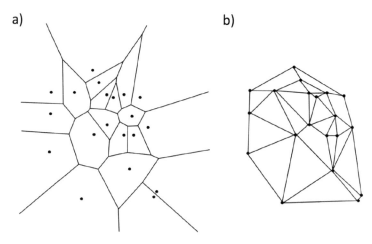

Figure 4.3.: a) Voronoi diagram and Delauny triangulation for $N = 20$ generator points. For a finite set of generator points $\mathbb{G} \subset \mathbb{G}^d$ the Voronoi diagram maps each $p \in \mathbb{G}$ onto its Voronoi region $R(p)$ consisting of all $x \in \mathbb{G}^d$ that are closer to p than to any other point in $\mathbb{G}$. b) Delaunay triangulation for the sites in a).

fields for describing space-filling, mosaic-like structures resulting from growth processes. Voronoi diagrams are geometric structures that deal with proximity of a set of points or more general objects. Often one wants to know details about proximity: Who is closest to whom? who is furthest and so on. The origin of this concept dates back to the 17th century. For example, in his book *on the principles of philosophy*, R. Descartes claims that the solar system consists of vortices. His illustrations show a decomposition of space into convex regions, each consisting of matter revolving round one of the fixed stars. Even though Descartes has not explicitly defined the extension of these regions, the underlying idea seems to be the following: Let a space $\mathbb{G}$ and a set S of sites p in $\mathbb{G}$ be given, together with the notion of the *influence* a site p exerts on a point x of $\mathbb{G}$. Then the region of p consists of all points x for which the influence of p is the strongest, over all $t \in \mathbb{G}$. This concept has independently emerged, and proven useful, in various fields of science. Different names particular to the respective field have been used, such as *medial axis transform* in biology or physiology, *Wiegner-Seitz zones* in chemistry and physics, *domains of action* in crystallography, and *Thiessen polygons* in meteorology. The mathematicians Dirichlet (1850) [57], and Voronoi (1908) were the first to formally introduce this concept. They used it for the study of quadratic forms; here the sites are integer lattice points, and influence is measured by the Euclidean distance. The resulting structure has been called *Dirichlet tesselation* or *Voronoi diagram*, cf. Fig. 4.3 a), which has become its standard name today. Voronoi was the first to consider the dual of this structure, where any two point sites are connected whose regions have a boundary in common, cf. Fig. 4.3 b). Later, Delauney obtained the same by defining that two point sites are connected if (and only if) they lie on a circle whose interior contains no point of $\mathbb{G}$. After him, the dual of the Voronoi diagram has been named *Delaunay tesselation* or *Delaunay triangulation*.

Voronoi tessellations in $\mathbb{R}^2$ have been used in many fields of materials science, e. g. for the description of biological tissues or polymer foams [226]. Ghosh et al. [86] utilized Voronoi cells to obtain stereologic information for the different morphologies of grains in ceramics and Espinoza et al. [72] used random Voronoi tessellations for the study of wave propagation models that describe various mechanisms of dynamic material failure at the microscale. However, these models have major drawbacks such as limitations to two dimensions and a generic nature of the structures as they are usually not validated with actual experimental data. Besides its applications in other fields of science, the Voronoi diagram and its dual can be used for solving numerous, and surprisingly different, geometric problems. A more or less complete overview over the existing literature can be found in the monograph by Okabe et al. [174] who lists more than 600 references, and in the survey by Aurenhammer [12].

In our approach to microstructural modeling of polycrystalline solids we use power diagrams (PDs) along with a new optimization scheme for the generation of realistic 3D structures [133]. PDs are a well studied generalization of Voronoi diagrams for arbitrary dimensions [174] and

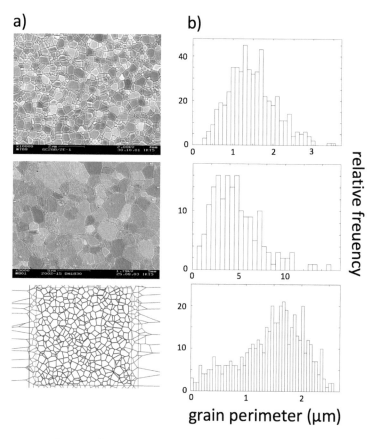

Figure 4.4.: Two different Al_2O_3 micrographs along with the 2D virtual cut through a non-optimized PD are shown in a) and their corresponding grain statistics with respect to the grains' perimeter are displayed under b). One can clearly notice that the histograms of the two experimental 2D grain structures do not follow a Gaussian distribution as was claimed, e.g. by Zhang et al. [256].

have some major advantages over Voronoi diagrams as outlined in [133]. The suggested optimization is based on the statistical characterization of the grains in terms of the distribution of the grain areas A and the grain perimeters P obtained from cross-section micrographs, cf. Fig. 4.4. An important result obtained using this method is that neither the experimental area nor the perimeter distribution obey a Gaussian statistics which is contrary to what was claimed e.g. by Zhang et al. [133, 256].

Our optimization scheme for the generation of realistic polycrystalline structures in 3D is based on comparing all polyhedral cells (typically at least 10,000) inside a cube of a given PD in 3D with the 2D experimental data. This comparison is performed for each coordinate axis by generating a set of parallel, equidistant 2D slices (typically 500 slices for each of the three coordinate directions) through the cube and perpendicular to the respective axis, see Fig. 4.5. For each 2D slice the grain sizes A are calculated and combined into one histogram. The same is done for the perimeter P. Then, the calculated histograms are compared with the experimental histograms A_i^{exp} and P_i^{exp} by calculating the first k central moments of the area and perimeter distributions A_i and P_i, respectively. A figure of merit m of conformity is defined according to which the PDs are optimized [133]:

$$m = \sum_{i=1}^{k} \left(\frac{P_i - P_i^{\mathrm{exp}}}{P_i^{\mathrm{exp}}} \right)^2 + \left(\frac{A_i - A_i^{\mathrm{exp}}}{A_i^{\mathrm{exp}}} \right)^2. \tag{4.1}$$

The figure of merit m in (4.1) is first calculated from the initial PD generated by a Poisson distribution of generator points. As a next step, using a reverse Monte-Carlo scheme, one generator point is chosen at random, its position modified and m is calculated and checked again. If m has decreased, the MC move is accepted, otherwise it is rejected. The modification of generator points is continued until m has reached a given threshold, typically 10^{-1}. If $m = 0$ is reached, the first k central moments of the experimental distributions agree completely with the model.

In Fig. 4.6 we present as an example the resulting histogram of an optimized PD for the Al_2O_3 material specimen of Fig. 4.1 a) together with the evolution of the figure of merit m for this sample during the reverse MC optimization procedure. After $358,000$ and

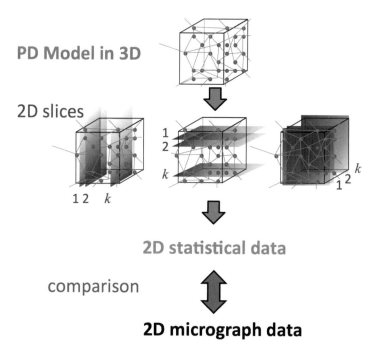

PD Model in 3D

2D slices

2D statistical data

comparison

2D micrograph data

Figure 4.5.: Our optimization scheme as suggested in [133]: A comparison is done between 2D experimental data and 2D slices obtained from the PDs in 3D as shown in Figs. 4.1 and 4.4.

$512,000$ optimization steps, the maximum step size of the reverse MC algorithm (changing the position of a generator point) was increased, which shows a direct influence on the speed of optimization. After 1.5 million optimization steps the deviation between the model and experiment has dropped below 1.3×10^{-4}.

Figure 4.7 shows a 3D tile of a meshed PD. In a continuum approach the considered grain structure of the material is typically subdivided into smaller (finite) elements, e.g. triangles in 2D or tetrahedra in 3D. Tetrahedral elements at the surface can either be cut, thus obtaining a smooth surface, or they can represent (a more realistic) surface coarseness. Also displayed in Fig. 4.7 is an enlarged section of the 3D

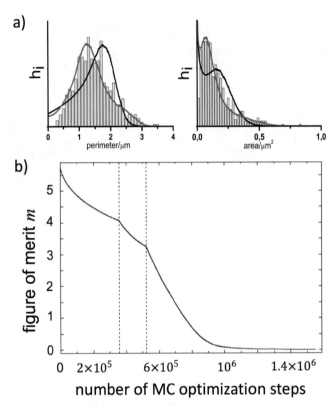

Figure 4.6.: a) Area (left) and perimeter (right) distribution of a PD
with 10,000 generator points before (dark curve) and after
(light curve) MC optimization. The bar graphs show the
respective histograms of the 2D experimental data of the
Al_2O_3 micrograph of Fig. 4.1. The quantity h_i along the
vertical axes indicates the relative frequency of grains with
a certain area/perimeter size. b) Evolution of the figure of
merit m during the optimization process.

tetrahedral mesh at the surface of the virtual specimen. Upon failure,
the elements are separated according to some predefined failure modes,
often including a heuristic Weibull distribution [242, 255] which is
artificially imposed upon the system.

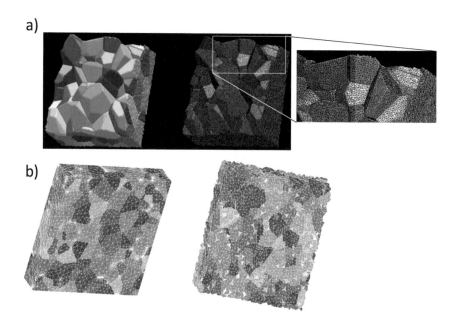

Figure 4.7.: Meshed 3D microstructure of an optimized PD generated
with the author's tool POLYGRAIN. In a) the granular surface
structure, its mesh and a detailed augmented section of the
mesh on the surface are displayed. b) A different realization
of a 3D structure displaying the possibility of either leaving
a (more realistic) rough surface microstructure, or smooth-
ing the surface and thus obtaining a model body with even
surface [133].

Having an efficient means to generate realistic polycrystalline structures, they can be meshed and be used for a numerical FE analysis. For
simulations of macroscopic material behavior, techniques based on a
continuum approximation, such as FEM or SPH are almost exclusively
used.

Accordingly, in Fig. 4.8, we present a snapshot from simulations
that explicitly mimic the situation of the experimental edge-on impact
setup discussed in Sec. 4.2. From these simulations, we obtain a
crack front velocity of the opening cracks in the numerical model of

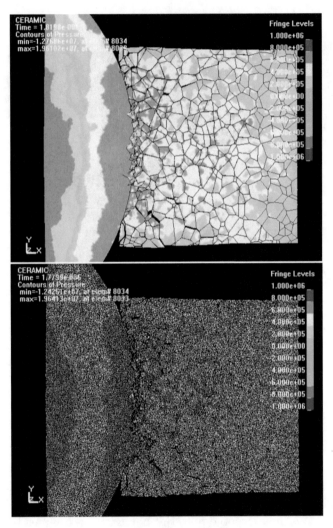

Figure 4.8.: Simulation of a steel sphere, impacting a thin ceramic plate
modeled with $1,452$ separate grains with "tiebreak contact"
and 1.2×10^6 finite elements. These simulations are done
in 3D with the engineering tool LS-DYNA. a) Granular structure with pressure levels between 12MPa (red) and 19MPa
(dark blue); b) Pressure levels with FEM mesh. Snapshots
are © Martin. O. Steinhauser.

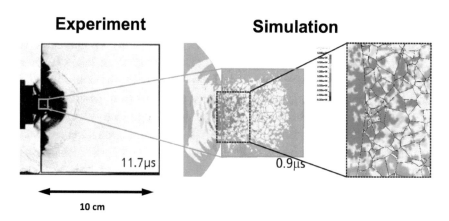

Figure 4.9.: Illustration of the multiscale problem. With concurrent FE methods which include microstructural details, only a very small part of a macroscopic system can actually be simulated due to the necessary large number of elements. For further details, see main text.

$v_{\text{crack}} \approx 4,900 \text{ms}^{-1}$, which is close to the results of our high-speed impact experiments, where we measure $v_{\text{crack}} \approx (4,700 \pm 230) \text{ms}^{-1}$.

Finally, in Fig. 4.9 we illustrate the disadvantages and the multiscale problem associated with FE simulations in which microstructural details are included. On the left, a high-speed camera snapshot of an edge-on impact experiment $11.7 \mu s$ after impact is shown, where a macroscopic steel sphere impacts the edge of an Aluminum Oxinitride (AlON) ceramic tile of dimension $(10 \times 10 \times 1)\,cm$. The enlargements in the middle and on the right show the small size of the region that is actually accessible to FE analysis in a concurrent multiscale simulation approach. With FE only a very small part of a macroscopic system can actually be simulated due to the necessary large number of elements. This is why in FE simulations of polycrystalline materials, in order to be able to simulate a sufficient number of grains, often only two dimensions are considered in the first place. For most codes, an element number exceeding a few dozen millions is the upper limit which is still feasible in FE simulations on the largest super computer systems.

More severe, the constitutive equations for the material description
which are needed in a phenomenological description, are derived from
experiments with idealized load conditions. This often leads to many
fit parameters in models, which diminishes their physical value. In
addition, the FEM generally has many computational problems (numerical instabilities) when it comes to very large element distortions
in the vicinity of the impact region where the stresses, strain rates,
and deformations are very large. The time scale of a multiscale FE
simulation does not a priori fit to the timescale of the experiment; thus,
parameter adaptations of the included damage model are necessary
(but are often totally unphysical). Also, we found that the contact algorithms implemented in common engineering codes such as pamcrash
or LS-DYNA which ensure that elements cannot penetrate each other in
impact situations, where high strain rates occur, are often unphysical,
very inefficient, and thus not well suited for parallelized applications.
The multiscale problem associated with FE simulations described in
Fig. 4.9 is further worsened by the fact that the results of FE analyses
of highly dynamic processes are often strongly influenced by mesh
resolution and mesh quality [51, 248] which, from a physical point
of view, is not acceptable, since the physical properties of a system
should be invariant to the arbitrarily chosen spatial resolution of the
problem.

4.2. High-speed impact experiments in solids

Experiments for high-speed loads of materials are often done in a
standard set-up, the edge-on impact (EOI) test, which allows to obtain reproducible results. In this standard experimental configuration,
the projectile hits an edge of the target specimen and the resulting
fracture propagation and final comminution is observed by means of a
high-speed camera system, see Fig. 4.10. In our modeling approaches
we explicitly simulate the projectile/sample interaction of this experimental set-up. In the velocity range up to 400ms^{-1} a gas gun using
compressed air is being employed for the acceleration of the projectiles.

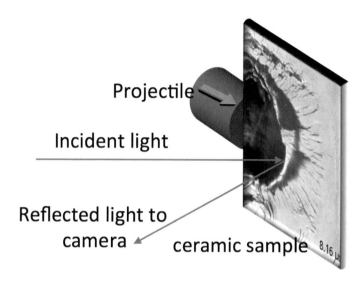

Figure 4.10.: Projectile/sample interaction in the EOI test configuration in a reflected light mode.

The ceramic specimens are placed at a distance of 1 cm in front of the muzzle of the gas gun in order to achieve reproducible impact conditions and a high precision measurement. In this set-up the rear of the projectile is still guided by the barrel of the gun when the front hits the target. The projectile velocity is measured by means of two light barriers through openings in the barrel 15cm behind the muzzle. For higher impact velocities of up to about $1,000\text{ms}^{-1}$, a 30mm caliber powder gun is used for the acceleration of the projectiles. In this case, due to the muzzle flash and the fumes, the specimens cannot be placed directly in front of the muzzle.

The Cranz-Schardin camera employed in our tests uses 24 sparks to expose the same number of images on a single sheet of photographic film. The sparks are distributed in 4 columns of 6 sparks and the light from the high-voltage discharges is imaged to an array of 24 objective lenses, arranged in the same geometry as the spark light sources. The

Figure 4.11.: Selection of 5 high-speed photographs from impact of a
SiC ceramic at striking velocity $v_s = 1,040\,ms^{-1}$.

imaging by means of a field lens causes the separation of the images on
the sheet film. The camera has no shutter, which means that the tests
have to be conducted in darkness. The frame rate is controlled by the
sequence of the light flashes. Each flash has a duration of about 250ns,
which also defines the exposure time. The maximum frame rate of the
high-speed spark unit is 10MHz. The high frame rate available allows
an optimal distribution of frames during the time interval of interest.

When the projectile hits the edge of the ceramic tile a shock wave
is generated which propagates through the material. For example,
in SiC the longitudinal wave speed is typically in the range from
12 to 13kms^{-1}. Hence, the wave will arrive at the rear edge of the
specimen after 7.5μs to 8.5μs. With the spark unit developed at EMI,
all pictures can be taken during this time interval. In general, in order
to visualize cracks, the surface of the specimen has to be polished
mirror-like, because otherwise, the intensity of the reflected light is
not sufficient. With Al$_2$O$_3$–ceramics a mirroring surface cannot be
accomplished by polishing only. In this case an additional coating
with a thin layer of silver or aluminum is required. At the positions
where the surface is disturbed by the crack edges, no light is directly
reflected to the camera. Thus, fracture appears as dark lines on the
photograph. The typical dimensions of the ceramic plates used are
$(100 \times 100 \times 10)$mm^3. Steel cylinders of 30mm diameter and 23mm
length are employed as projectiles with impact velocities in the range
from 20ms^{-1} to $1,100$ms^{-1}.

In Fig. 4.11 we display a selection of 5 high-speed photographs of a SiC ceramic impacted at $1,040\text{m/s}$. A semicircular shaped fracture front can be recognized, which follows closely behind the longitudinal pressure wave. The analysis of the photographs yields a wave velocity of 12.59kms^{-1} and a propagation velocity of the fracture front of 11.15kms^{-1}. Different fracture patterns in different materials occur in the EOI experiments due to different mechanical characteristics of the materials such as different densities, mesoscopic grain sizes, or hardnesses. In addition, the kind of observed fracture patterns change when using different geometries for the impactors, e.g. spheres instead of a cylindrical shape have demonstrated that the observed variations of failure behavior in different brittle materials, e.g. the degree of damage as a function of impact velocity, can be replicated by means of non-equilibrium MD simulations.

4.3. DEM modeling of shock wave failure in granular materials

In this section we discuss a concurrent multiscale approach for the simulation of failure and cracks in brittle materials which is based on mesoscopic particle dynamics, the discrete element method (DEM), but which allows for simulating macroscopic properties of solids by fitting only a few model parameters [215]. Instead of trying to reproduce the geometrical shape of grains on the microscale as seen in 2D micrographs, in the proposed approach one models the macroscopic solid state with soft particles, which, in the initial configuration, are allowed to overlap, cf. Fig. 4.12 a). Here, only two particles are displayed; the number of overlapping particles however, is unlimited and *each* individual particle pair contributes to the overall pressure and tensile strength of the solid. The overall system configuration, see Fig. 4.12 b), can be visualized as a network of links that connect the centers of overlapping particles, cf. Fig. 4.12 c).

The degree of particle overlap in the model is a measure of the force that is needed to detach particles from each other. The force is imposed on the particles by elastic springs. This simple model can easily be extended to incorporate irreversible changes of state such as plastic flow in metals on the . However, for brittle materials, where catastrophic failure occurs after a short elastic strain, in general, plastic flow behavior can be completely neglected. Additionally, a failure threshold is introduced for both, extension and compression of the springs that connect the initial particle network. By adjusting only two model parameters for the strain part of the potential, the correct stress-strain relationship of a specific brittle material as observed in (macroscopic) experiments can be obtained. The model is then applied to other types of external loading, e.g. shear and high-speed impact, with no further model adjustments, and the results are compared with high-speed impact experiments performed at EMI.

4.3.1. Interaction potentials

The main features of a coarse-grained model – in the spirit of Occam's razor with only few parameters – are the repulsive forces which determine the materials resistance against pressure and the cohesive forces that keep the material together. A material resistance against pressure load is introduced by a simple Lennard Jones type repulsive potential ϕ_{rep}^{ij} which acts on every pair of particles $\{ij\}$ once the degree of overlap d^{ij} decreases compared to the initial overlap $d_0{}^{ij}$:

$$
\phi_{\text{rep}}^{ij} \left(\gamma, d^{ij} \right) = \begin{cases} \gamma R_0{}^3 \left(\left(\frac{d_0{}^{ij}}{d^{ij}} \right)^{12} - 2 \left(\frac{d_0{}^{ij}}{d^{ij}} \right)^6 + 1 \right) & 0 < d^{ij} < d_0{}^{ij} \\ 0 & d^{ij} \geq d_0{}^{ij} \end{cases}.
$$

(4.2)

Parameter γ scales the energy density of the potential and prefactor $R_0{}^3$ ensures the correct scaling behavior of the calculated total stress $\Sigma_{ij}\sigma^{ij} = \Sigma_{ij}F^{ij}/A$ which, as a result, is independent of N. Our mathematical proof of this can be found in [215]. Figure 4.13 shows the scaling property of our model: The original system (Model M_a) with edge length L_0 and particle radii R_0 is downscaled by a factor of $1/a$ into the subsystem Q_A of M_A (shaded area) with edge length L, while the particle radii are upscaled by factor a. As a result, model

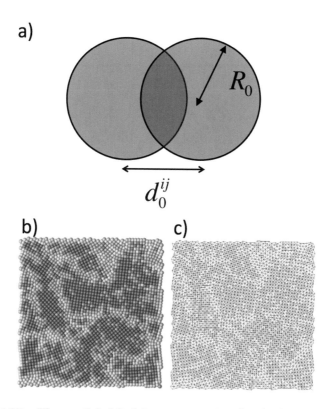

Figure 4.12.: The particle Model as suggested in [215]. a) Overlapping
particles with radii R_0 and the initial (randomly generated) degree of overlap indicated by $d_0{}^{ij}$. b) Sample
initial configuration of overlapping particles as 2D cut
($N = 25,000$) where dark color indicates high particle
density. c) The same system displayed as a network of
bonds.

M_B of size $aL = L_0$ is obtained containing much fewer particles,
but representing the same macroscopic solid, since the stress-strain
relation (and hence, Young's modulus E) upon uni-axial tensile load
is the same in both models. Hence, systems with all parameters
kept constant, but only N varied, lead to the same slope (Young's

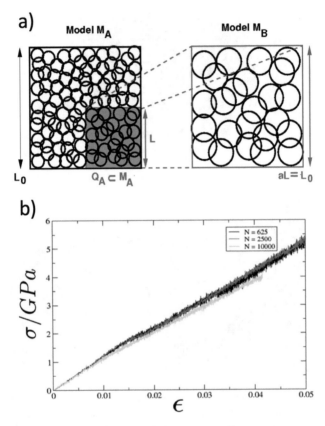

Figure 4.13.: a) Schematic of the intrinsic scaling property of the proposed material model. Here, only the 2D case is shown for
simplicity. b) Young's modulus E of systems with different
number of particles N in a stress-strain ($\sigma - \epsilon$) diagram. In
essence, E is indeed independent of N.

modulus) in a stress-strain diagram. In (4.2) R_0 is the constant radius
of the particles, $d^{ij} = d^{ij}(t)$ is the instantaneous mutual distance of
each interacting pair $\{ij\}$ of particles, and $d_0^{\,ij} = d^{ij}(t = 0)$ is the
initial separation which the pair $\{ij\}$ had in the starting configuration.
Every single pair $\{ij\}$ of overlapping particles is associated with a

different initial separation $d_0{}^{ij}$ and hence with a different force. The minimum of each individual particle pair $\{ij\}$ is chosen such that the body is force-free at the start of the simulation. When the material is exposed to tensile load, the small deviations of particle positions from equilibrium will vanish again, as soon as the external force is released. Each individual pair of overlapping particles can thus be visualized as being connected by a spring, the equilibrium length of which equals the initial distance $d_0{}^{ij}$. This property is expressed in the cohesive potential by the following equation:

$$\phi_{\text{coh}}^{ij}\left(\lambda, d^{ij}\right) = \lambda R_0 \left(d^{ij} - d_0{}^{ij}\right)^2, \ d^{ij} > 0. \tag{4.3}$$

In this equation, λ (which has dimension [energy/length]) determines the strength of the potential and prefactor R_0 again ensures a proper intrinsic scaling behavior of the material response. The total potential is the following sum:

$$\phi_{\text{tot}} = \Sigma_{ij} \left(\phi_{\text{rep}}^{ij} + \phi_{\text{coh}}^{ij}\right). \tag{4.4}$$

The repulsive part of ϕ_{tot} acts only on particle pairs that are closer together than their mutual initial distance $d_0{}^{ij}$, whereas the harmonic potential ϕ_{coh} either acts repulsively or cohesively, depending on the actual distance d^{ij}. Failure is included in the model by introducing two breaking thresholds for the springs with respect to compressive and to tensile failure, respectively. If either of these thresholds is exceeded, the respective spring is considered to be broken and is removed from the system. A tensile failure criterium is reached when the overlap between two particles vanishes, i.e. when:

$$d^{ij} > (2R_0). \tag{4.5}$$

Failure under pressure load occurs when the actual mutual particle distance is less by a factor α (with $\alpha \in (0,1)$) than the initial mutual particle distance, i.e. when

$$d^{ij} < \alpha \cdot d_0{}^{ij}. \tag{4.6}$$

Particle pairs without a spring after pressure or tensile failure still interact via the repulsive potential and cannot move through each other.

An appealing feature of this model, as opposed to many other material models used for the description of brittle materials, see e.g. [90, 115, 142, 194, 235], is its simplicity. The proposed model has a total of only three free parameters: γ and λ for the interaction potentials and α for failure. These model parameters can be adjusted to mimic the behavior of specific materials. The initial particle structure of the system is generated in a warmup process before the actual simulation, during which the simulation box is filled with particles which are allowed to grow for an expansion time τ, until the desired particle radii R_0 and overall particle density Θ are reached.

The expansion time τ during the warmup procedure and the preset particle density Θ of the system determine the structure of the initial particle configuration. In order to be able to specifically tune these parameters such that the global behavior of the material corresponds to experimental data, we first investigated the influence of these parameters on the resulting breaking strength σ_b in a uniaxial strain simulation. From the initial particle pair distance distribution $\langle d_0{}^{ij} \rangle$ one can derive a maximal expectation value for σ_b:

$$\sigma_b^{\text{max, theor.}} = 1 - \frac{\langle d_0{}^{ij} \rangle}{2R_0}, \tag{4.7}$$

where R_0 is the constant radius of all particles. The maximum tensile strength does not appear as a parameter in the model; it follows from the initial particle configuration and the resulting initial density Θ, see Fig. 4.14. The actual breaking strength of the material is much smaller than the theoretical strength value since our system is disordered. This results in a faster failure upon tensile load. The longer the expansion time τ, the more ordered is the resulting initial structure with a corresponding reduced variance in the pair distance distribution. A more ordered structure results in a higher tensile strength as there are less weak connections with particle distances that are already close to the failure threshold $2R_0$.

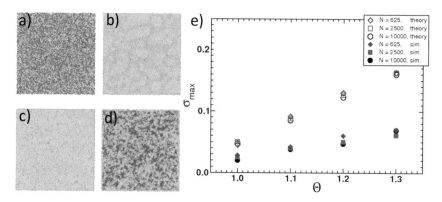

Figure 4.14.: Analysis of generated DEM material samples. Sample
initial configurations of systems with $N = 10.000$ particles
and different particle densities Θ: a) $\Theta = 0.7$; b) $\Theta = 0.9$;
c) $\Theta = 1.3$; d) $\Theta = 1.7$; e); Breaking strength σ_b for
different system sizes N (filled symbols) as a function of
particle density Θ, compared with the theoretical breaking
strength σ_{max} (open symbols).

The maximum tensile strength does not appear as a parameter in
the model; it follows from the initial particle configuration and the
resulting initial density Θ, see Fig. 4.14. The closer $\langle d_0^{ij} \rangle$ is to $2R_0$,
the less is the maximum tensile strength. The random distribution of
initial particle distances ultimately determines the system's stability
upon load, as well as its failure behavior and the initiation of cracks.

4.3.2. Starting configurations of the DEM model

A simple and efficient way of modeling a granular polycrystal material
is the use of mono-disperse spheres in 3D. Using polydisperse particle
distributions directly influences characteristic properties of the material, e.g. the coordination number of the starting configuration. In our
simple approach we use mono-disperse spheres and adjust the initial
coordination number of the system via a compactness parameter Θ as
simulation input.

The physical observables of a given solid are, of course, independent of the number of particles chosen for its discretization. In the same way, the corresponding mesoscopic properties should be independent of N. Therefore, in our approach, the main properties of the material to be modeled enter through the specific potentials between the particles.

In our study presented here, we use as initial configuration a random distribution of particles in a cubic simulation box, on which we impose periodic boundary conditions. The disks are allowed to overlap. The degree of overlap determines the force that is needed to separate the disks upon tension or to further squeeze the system upon pressure load. The initial radii $R_0(\Theta)$ of the disks are determined by the equation:

$$R_0(\Theta) = \sqrt{\frac{L_{Box}^3 \Theta}{N\pi}}, \qquad (4.8)$$

where L_{Box} is the linear dimension of the simulation box, N is the chosen number of particles, and Θ is a dimensionless compactness parameter which is related to the actual density $F(\Theta)$ of particles, i.e. the ratio of the volume which is filled with matter and the total volume. Thus, by tuning the compactness parameter, we can fix the density of the material in our model according to the one obtained in sintered ceramics of interest, e.g. $\rho \geq 98\%$ in the case of Al_2O_3 and SiC.

If the MD simulation was started with the full strength of the potentials, the system would be numerically unstable due to very large forces between strongly overlapping particles. Therefore, in a first step for generating a random starting configuration, the particles' radii are reduced from the fixed preset size of (4.8), such that none of them overlaps, i.e. all particles are disjoint.

In a second step, during a MD warm-up phase the particle radii are gradually increased (typically during $\Delta t = 20,000$ time steps) until they reach their final value R_0. The expansion time τ of the particles is proportional to the number of integration steps Δt that are used for successively expanding them from their initial non-overlapping size to the final size R_0. With every expansion step the system is driven out of equilibrium. The slower this process occurs, i.e. the more integration steps are used for this procedure, the closer the system evolves to an equilibrium state.

We end the warm-up procedure as soon as the difference of the potential energy of two system configurations, which are Δt time steps apart, is smaller than 1%. The basic structure of the resulting particular initial particle configuration is determined by the overall density in the system, fixed by Θ and by the expansion time τ.

In order to derive an analytic formula for the particle density in 2D, we consider in the following a completely regular hexagonal arrangement of disks. Varying Θ, one identifies three regimes for the fraction $F(\Theta)$ of the area covered by the disks. Based on simple geometric considerations, one obtains:

$$F(\Theta) = \begin{cases} \Theta & : (0 \le \Theta < \Theta_1) \\ \Theta\{1 - \frac{3}{\pi}[G(\Theta) - \sin G(\Theta)]\}: & (\Theta_1 \le \Theta < \Theta_2) \\ 1 & : (\Theta_2 \le \Theta) \end{cases} . \quad (4.9)$$

The first domain is the one in which none of the disks overlap, and is limited by $\Theta_1 = \pi/(2\sqrt{3}) \approx 0.9069$. For values of Θ getting larger than Θ_1, pairs of disks begin to overlap. When further increasing the value of Θ one reaches another critical value (Θ_2 in (4.9)) from which on groups of three disks overlap. In this case the area of the simulation box is completely covered and $F = 1$, independent of a further increase of Θ. The value of $\Theta_2 = 4/3 \cdot \Theta_1 = 2\pi/\sqrt{27} \approx 1.2092$. For the intermediate domain the fraction of the covered area is given by the second expression in (4.9), where $G(\Theta)$ is:

$$G(\Theta) = 2\arccos\left(\sqrt{\frac{\Theta_1}{\Theta}}\right). \quad (4.10)$$

We can use the difference between the value of $F(\Theta)$ determined numerically from computer generated configurations and the analytic value given by (4.9) as a measure of the degree of inhomogeneity of our random configurations. We observe that the generated systems using our model gradually approach an ordered structure with increasing

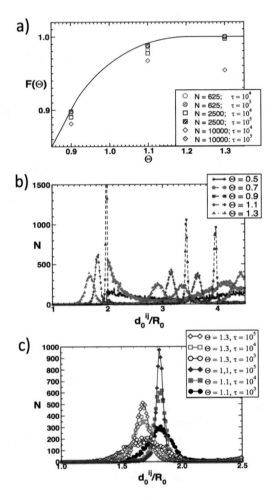

Figure 4.15.: a) Analytic curve $F(\Theta)$ and simulation data for different
system sizes N and different expansion times τ. b) Distribution of mutual distances d^{ij}/R_0 for different values of
Θ in a system with $N = 2,500$ particles. c) Distribution
of mutual distances d_0^{ij}/R_0 for two different Θ-values and
several expansion times τ in a system with $N = 2,500$. All
other system parameters are kept constant. The expansion time influences the width of the distribution and Θ
determines the location of the mean value.

compactness Θ and increasing expansion time τ, cf. Fig. 4.15. In Fig. 4.15 a) we have plotted (4.9) in comparison with $F(\Theta)$, determined for several initial configurations with different Θ-values, different system sizes N and different expansion times τ.

Figure 4.15 b) displays for different Θ the nearest and next-nearest neighbor distance distributions. With increasing Θ, the peaks in the pair distributions become more pronounced, which indicates a more regular arrangement of particles.

Figure 4.15 c) finally exhibits the pair distance distribution for a system with $N = 2,500$ particles for two different Θ-values and various expansion times τ. The distribution becomes sharper with increasing τ and the average value $\langle d_0{}^{ij} \rangle$ decreases with increasing Θ.

4.3.3. Results and comparison with experiments

In this section we present quasi-static load simulations using our model in order to fix the model parameters to specific material values of Young's modulus and the breaking strength. After adjusting the parameters we test our model by measuring the Poisson number and the shear modulus G in uni-axial tensile load simulations for Al_2O_3 and obtain good agreement with experiments. the system is sheared and an impact experiment, as shown in Figs. 4.10 and 4.11 is performed. Finally, without any further fitting of parameters, we present impact simulations with our model in situations with high strain-rates and compare these results with our corresponding high-speed EOI experiments described in Sec. 4.2.

Quasi-static simulations and crack initiation

In quasi-static load experiments, the involved physical processes occur on time scales small enough so that the system under investigation is always very close to equilibrium. We strain the material instantan-eously on two opposite sides and then allow it to relax by means of a molecular dynamics integration scheme with a time step $\Delta t = 0.001\tau$ in reduced simulation units. During relaxation the particles may

move and bonds may break. Equilibrium is reached per definition as
soon as no bonds break during $1,000$ consecutive time steps. After
reaching equilibrium, the external strain is increased again and the
whole procedure is iterated until ultimate failure occurs. For the actual
simulation adapted to Al_2O_3, $\alpha = 0.99$, $\gamma = 10.0$, $\lambda = 350$ and $\Theta = 1.1$
are chosen; these values correspond to a typical experimental situation
of 99% volume density, $\rho = 3.96\text{g/cm}^3$ and $E = 370\text{GPa}$.

Results of the simulations are displayed in the picture series of
Fig. 4.16 a)-d), which shows 4 snapshots of the fracture process, in
which the main features of crack instability, as pointed out by Fineberg [76] (onset of branching at crack tips, followed by crack branching
and fragmentation) are well captured. At first, many micro-cracks
are initiated all over the material by failing particle pair bonds, cf.
Fig. 4.16 a). These micro-cracks lead to local accumulations of stresses
in the material until a macroscopic crack originates (Fig. 4.16 b), which
quickly leads to a complete rupture of the macroscopic sample, see
Figs. 4.16 c) and d).

A characteristic material parameter which can be determined from
tensile experiments is Young's modulus E. E is simply the slope of
a stress-strain curve obtained from a tensile load experiment. From
analyzing simulated stress-strain relations for different values of κ
which determines the cohesive strength of the particle interactions, we
obtain the following result:

$$\frac{E}{\kappa} = (1.058 \pm 0.013) \left[\frac{10^9}{m}\right]. \qquad (4.11)$$

Hence, for a given modulus, (4.11) fixes the value of the cohesive
potential parameter κ. For Al_2O_3, we obtain a Poisson number of

$$\mu = 0.222 \pm 0.015, \qquad (4.12)$$

which agrees well with the experimental values of around $\mu = 0.22$.

Without any further model adjustments we next perform a shear-
load simulation, the results of which are presented in Fig. 4.17. Only
the color-coded network of particles is shown: stress-free (green),

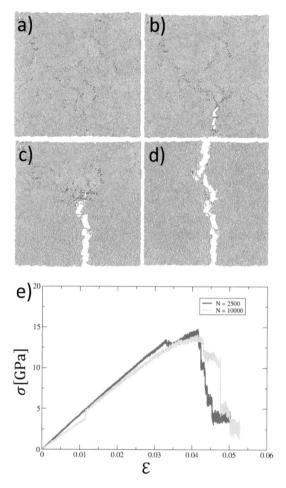

Figure 4.16.: Crack initiation and propagation in the virtual macroscopic material sample upon uni-axial tensile load using
$N = 10^4$ particles. a) Initiation of local tensions. b) Initiation of a crack tip with local tensions concentrated
around this crack tip. c) Crack propagation including
crack instability. d) Failure. e) Averaged stress-strain (σ–ϵ)
relation. For $N = 2,500$, ten different systems were av-
eraged, and for $N = 10,000$ the average for five different
systems is displayed.

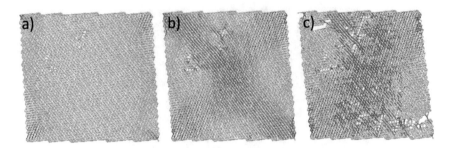

Figure 4.17.: Quasi-static shear load of a virtual material specimen
with $N = 2,500$. a) Onset of shear tensile bands and
(orthogonal) shear pressure bands in the corners of the
specimen. b) Shear bands traversing the whole specimen.
c) Ultimate failure.

tension (red) and pressure (blue). Starting from the initially unloaded
state, the top and bottom layer of particles is shifted in opposite
directions. At first, in Fig. 4.17 a), the tension increases over the
whole system. Then, as can be seen from Fig. 4.17 b) and c), shear
bands form and stresses accumulate, until failure due to a series of
broken bonds occurs. From a total of 60 different computer simulations
we determine the average shear modulus G for Al_2O_3 to be

$$\langle G \rangle = (120 \pm 3) GPa, \qquad (4.13)$$

a value only about 15% below typical value found for Al_2O_3. The
shear modulus is obtained from the slope of the curve in a shear stress
τ vs. shear strain γ diagram. In light of the fact that in our model is
very simple with only three free parameters we find our results, (4.12)
and (4.13), rather remarkable.

Dynamic impact simulations

In this section we test our DEM model for the case of dynamic material
behavior where high strain rates occur. Numerically, this corresponds
to dynamic non-equilibrium simulations. For this we use the EOI1
configuration as described in the experimental section, Sec. 4.2.

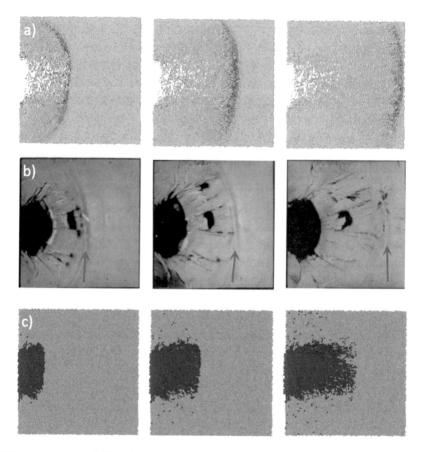

Figure 4.18.: a) Results of an EOI simulation at $v = 150$ m/s with SiC using $N = 10^5$ particles. The material is hit at the left edge. A pressure wave (high pressure: dark regions, unloaded material: bright regions) propagates through the system. The time interval between the individual snapshots from left to right is 2μs in each case. b) The EOI experiment with a SiC specimen. The time interval between the photographs is comparable to that in the top row. The arrows indicate the location of the wave front propagating in the material. c) The same computer simulation as in a)-c), now displaying the occurring damage in the material with respect to broken bonds.

As a timestep for our multiscale non-equilibrium simulations we use $\Delta t = 0.001\tau$ in reduced simulation units, where τ is given by $\tau = l\sqrt{m/\gamma}$. The parameter l determines the length scale of the system, m is the mass of one particle, and γ is the prefactor of the potential in (4.2). For Lennard-Jones type potentials, σ is simply the equilibrium distance of every particle pair, in our case $\sigma = \langle d_0^{ij} \rangle$ as the initial particle distances are randomly distributed. To mimic the characteristic behavior of these materials we adapt the free parameters of our model such that we obtain from simulations the correct densities, Young's moduli, breaking strength upon tension and pressure and also the correct longitudinal wave velocities for Al_2O_3 and for SiC, respectively. For our model we showed that the tuning of parameters to specific materials is sensitive enough to capture differences in the degree of damage for different materials such as SiC or Al_2O_3 [216].

As we do not model the impactor explicitly, we transform its total kinetic energy into the kinetic energy of the particles in the impact region. Irreversible deformations of the particle network due to plasticity or heat are not considered in our model, i.e. energy is only removed from the system by broken bonds. Therefore, the development of damage in the material is slightly overestimated.

The top series of snapshots in Fig. 4.18 a) shows the propagation of a shock compression wave through the material. the distance and also the shape of the shock front correspond very well to the photographs in b) in the middle of Fig. 4.18. These snapshots were taken at comparable times after the impact had occurred in the experiment and in the simulation, respectively. As a wave velocity we obtain from our simulations roughly $v = 12,250\mathrm{m/s}$ which is in good agreement with the experiments

The advantage of simulations for investigating impact phenomena is evident from the simulation snapshots in Fig. 4.18. One can study in great detail the physics of shock waves traversing the material and easily identify strained or compressed regions in the material by plotting the potential energies of the individual bonds. Also the failure in the material can be conveniently visualized by plotting only the failed bonds as a function of time, cf. the bottom series

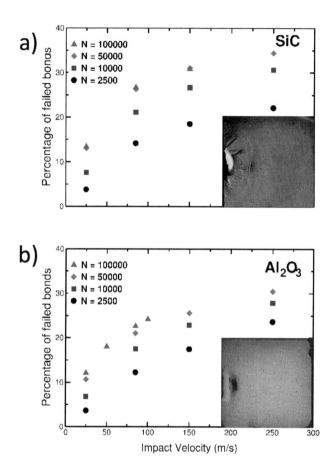

Figure 4.19.: Degree of damage at $3\mu s$ after impact for SiC and Al$_2$O$_3$
at different simulated impact velocities an different number
N of particles used in the model. The insets show high-
speed camera snapshots of SiC and Al$_2$O$_3$ indicating the
degree of damage and the advancing shock wave in the
experiment after impact with striking velocity $v = 85\text{m/s}$.
For system sizes of $N \approx 50,000$ and larger, finite-size
effects due to particles at the edges of the specimen begin
to cease, which is seen in the convergence of data points
for larger systems.

c) of snapshots in Fig. 4.18. These figures display the failed area in the material as a function of time. Analyzing the high-speed photographs in the picture series in Fig. 4.18 b) one obtains as crack velocity with which the main fracture front propagates through the specimen a value of roughly $v = 7,400\text{ms}^{-1}$ agrees with corresponding experimental measurements that yield a variation of crack velocities between $4,960\text{ms}^{-1}$ to $5,380\text{ms}^{-1}$ and isolated cracks propagating at a speed of $7,950\text{ms}^{-1}$. Note that our model not only yields continuously growing cracks but also some isolated cracks in those areas of the material where the initial overlap of particles is smaller than in other areas. Hence, while the *average global* failure behavior of our model, e.g. with respect to tensile strength, is determined by fitting the corresponding model parameters to experimental values, the *local* failure behavior is determined by the *local* random initial configuration of the particles. After a reflection of the pressure wave at the free end of the material sample, and its propagation back into the material, the energy stored in the shock wave front finally disperses in the material without causing further damage, i.e. further broken bonds. Finally, in Fig. 4.19 we show a detailed analysis of broken bonds in two different high-performance ceramics. We see consistently a higher damage in SiC compared to Al_2O_3 which is consistent with experimental observations. However, in experiments it is very difficult to make quantitative measurements of the degree of damage in the material. It is here where the simulations can not only complement but extend the experiments.

In summary, our model reproduces the physics of shock wave propagation in brittle materials very well. Numerically, the impact simulations using DEM are very efficient and scale roughly as $N \log N$, as we do not have to make sure to process our system in time from one equilibrium state to the next one. The generation and propagation of shock waves leads to a non-equilibrium evolution of the considered system. Therefore, using molecular dynamics simulations with discrete elements (in our case, particles) turns out to be a very interesting method for investigating damage and failure in shock compression of materials.

Part III.

Soft matter

5. Coarse-grained modeling and simulation of macromolecules

5.1. What is coarse-graining?

Coarse-grained (CG) models were originally introduced in macromolecular modeling approaches for globular proteins in a 1976 pioneering paper by Warshel and Levitt [241] (then called hybrid classical/quantum mechanical approach). Since then, CG models have found their way into polymer physics as so-called bead-spring models, taking advantage of universal scaling laws of long polymer chains due to their fractal nature [64, 206, 224, 244], as well as into geophysics, engineering and other areas of computational research [215, 217].

CG models of macromolecules provide a route to explore biomolecular systems on larger length and time scales while still resolving important physical aspects of the lipid bilayer structure [87, 148, 152, 170, 176, 206, 209, 211]. They constitute a class of mesoscale models, in which many atoms or groups of atoms are treated by grouping them together into new particles which act as individual interaction sites usually connected by entropic springs, see Fig. 5.1. Such models are becoming increasingly popular [14, 228] because they remain particle-based while greatly reducing the computational expense. Another advantage of CG methods is that the particle groups can be constructed at various resolutions, permitting the study of membranes at multiple scales.

A "bottom-up" CG model is a model of a particular system that is constructed on the basis of a more detailed model for the same

© Springer Fachmedien Wiesbaden GmbH 2018
M. O. Steinhauser, *Multiscale Modeling and Simulation of Shock Wave-Induced Failure in Materials Science*,
https://doi.org/10.1007/978-3-658-21134-9_5

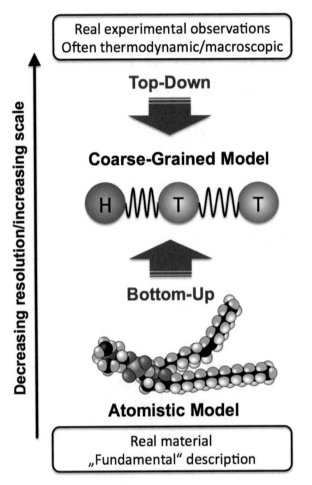

Figure 5.1.: Scheme illustrating top-down and bottom-up strategies for developing CG computational models for a common phospholipid molecule (DPPC, dipalmitoylphosphatidylcholine, $C_{40}H_{80}NO_8P$) most frequently occurring in plasma membranes [6] The CG model is a typical bead-spring model of polymer physics composed of three parts, one hydrophilic head (H) particle and two hydrophobic tail (T) particles, connected by bonds that are described by entropic springs. [211]. Figure taken from [223].

system as indicated in Fig. 5.1. In principle, the high-resolution, all-atom model may be based on atomistic data deduced from atomistic structure calculations.

In contrast, "top-down" models do not rely upon or directly relate to a more detailed model for a particular system. Instead, they are usually related to the full complexity of the real experimental system by addressing observables on length scales that are accessible to the CG model. Often, these observables are thermodynamic averaged quantities such as pressure, temperature, stress and strains or forces accessible by direct experimental measurement. Figure 5.1 illustrates schematically these two major approaches to coarse-graining.

CG simulations are much less computationally expensive than their atomistic counterparts, because the number of interacting particles is drastically reduced and can access much larger length- and time scales than is possible in all-atom approaches, let alone in quantum chemical calculations. [81, 170, 246] Coarse graining procedures may simply remove certain degrees of freedom (e.g. vibrational modes between two atoms) or it may in fact simplify the two atoms completely via a single particle representation. The ends to which systems may be coarse grained is simply bound by the accuracy in the dynamics and structural properties one wishes to replicate. The challenge of this modern area of research is still in its infancy, and although it is commonly used in biological modeling and polymer physics, the analytic theory behind it is still poorly understood.

Recent literature dealing with multiscale methods in polymer physics includes the edited volume *Multiscale Molecular Methods in Applied Chemistry* [119] and the treatment of multiscale computational methods in the monograph *Computational Multiscale Modeling of Fluids and Solids – Theory and Applications* [211]. Additional, very recent publications on this subject can be found at the website[1] of the 2016 special issue *Computational Multiscale Modeling and Simulation in Materials Science* of the Journal *Materials*, edited by Martin O. Steinhauser.

---

[1]http://www.mdpi.com/journal/materials/special_issues/

5.1.1. Coarse-graining of soft matter: polymers and biomacromolecules

The field of CG simulations of polymers and of biological macromolecular structures has seen an exciting development over the past 30 years, mostly due to the emergence of physical approaches in the biological sciences, which lead to the investigation of soft biomaterials, structures and diseases as well as to the development of new medical treatments and diagnostic methods [29, 236, 237, 239, 240]. The challenge faced here by material scientists, namely that structure and dynamics of materials are characterized by extremely broad spectra of length and time scales, is particularly apparent with polymeric materials. Due to their length, polymers can attain a large number of conformations which contribute to the total free energy of a macromolecular system. The structure of macromolecules is thus determined by an interplay between energetic and entropic contributions. The hydrodynamic interactions of polymers with solvent molecules, covalent interactions, intra- and intermolecular interactions and – particularly in biological macromolecules – the Coulomb interactions between charged monomers along with hydrogen bonds add up with the entropic forces to build a very complex system of interacting constituents. The enormous range of characteristic time and length scales accounts for their widespread use in technological applications and furthermore the most important biomolecular structures are polymers. It is clear that the treatment of such systems calls for hierarchical multiscale modeling approaches which can efficiently sample the complex potential energy hypersurface of polymeric materials, ensuring equilibration over the relevant length and time scales.

With polymer systems, many properties can be successfully simulated by only taking into account the basic and essential features of the chains, thus neglecting the detailed chemical structure of the molecules. Such *coarse grained models*, cf. Figure 5.2, are used because atomistic models of long polymer chains are usually intractable for time scales beyond nanoseconds, but which are important for many physical phenomena and for comparison with real experiments. Also,

Figure 5.2.: A coarse-grained model of a polymer chain where some groups of the detailed atomic structure are lumped into few coarse-grained particles connected by entropic springs (*bead-spring model*). Thus, information about the detailed chemical structure including the degree of freedom of bond angles is lost. The semiflexibility of a polymer on a large scale (expressed by the persistence length L_p) however, can be captured in CG models by introducing appropriate interactions, cf. [222].

the fractal properties of polymers render this type of model very useful for the investigation of scaling properties. Fractality means, that some property X of a system (with polymers, e.g. the radius of gyration R_g) obeys a relation $X \propto N^k$, where $N \in \mathbb{N}$ is the size of the system, and $k \in \mathbb{Q}$ is the fractal dimension which is of the form $k = p/q$ with $p \neq q$, $q \in \mathbb{N}$ and $p \in \mathbb{N}$. The basic properties that are sufficient to extract many structural static and dynamic features of polymers are:

- The connectivity of monomers in a chain.

- The topological constraints, e.g. the impenetrability of chain bonds.

- The flexibility or stiffness of monomer segments.

Using coarse-grained models in the context of lipids and proteins, where each amino acid of the protein is represented by two coarse-grained beads, it has become possible to simulate lipoprotein assemblies and protein-lipid complexes for several microseconds.

5.1.2. Crossover scaling of linear, semiflexible polymers

In this section we discuss our results of CG modeling of semiflexible polymers. Semiflexibility, i.e. stiffness of a polymer chain is a hallmark of all biological macromolecules, which is why we have to include it into our modeling schemes for biological macromolecular systems. We do this with an additional bending potential which provides an energetic penalty in case the bond angle – defined by the angle between consecutive segments connecting the monomers along the polymer chain – deviates from the preset angle. We show the correct implementation of semiflexibility of polymers in our simulations by analyzing their universal scaling properties when going from fully flexible behavior to semiflexible behavior. Since we employ CG models of polymers, we are able to simulate much longer persistence lengths L_p in our polymer systems that has ever been done in atomistic simulations of universal scaling properties. Atomistic simulations done by Paul et al. [34, 91, 180] claimed a scaling exponent in the range of $\approx p^{-2.3}$ to $\approx p^{-3.0}$, while theory predicts for the semiflexible regime a universal exponent of $p^{-4.0}$, where p is the Rouse mode number. In our simulations we could show for the first time this correct crossover scaling behavior in the transition from fully flexible to semiflexible behavior of linear macromolecules in computer simulations. We show our results for polymer melts composed of linear chain molecules. We then proceed to introduce our CG model for phospholipid molecules which we use for simulations of equilibrium properties of lipid bilayers and for shock wave simulations of membranes, which are an important part of eukariotic cells, determining their mechanical stability.

In contrast to theoretical modeling, only very few simulation studies using mostly detailed atomistic models have been performed on the behavior of semiflexible polymer systems [34, 128, 132, 180]. In particular the results of atomistic simulations have shown that the mean square particle displacements and the scattering functions deviate significantly from what is expected under ideal (Gaussian) conditions [180].

The Kratky-Porod chain model (or worm-like chain model) [127] provides a simple description of inextensible semiflexible polymers with positional fluctuations that are not purely entropic but governed by their bending energy U_{bend} and characterized e.g. by their persistence length L_p. The corresponding elastic energy

$$U_{\mathrm{Bend}} = \frac{\kappa}{2} \int_0^L ds \left(\frac{\partial^2 \vec{r}}{\partial s^2} \right)^2 \tag{5.1}$$

of the inextensible chain of length L depends on the local curvature of the chain contour s, where $\vec{r}(s)$ is the position vector of a mass point (a monomer) on the chain and κ is a constant [58].

Harris and Hearst formulated an equation of motion for the Kratky-Porod model by applying Hamilton's principle with the constraint that the second moment of the total chain length be fixed and obtained the following expressions for the bending $\vec{F}_{\mathrm{Bend}}$ and tension forces $\vec{F}_{\mathrm{Tens}}$

$$\vec{F}_{\mathrm{Bend}} = \kappa \frac{\partial^4 \vec{r}}{\partial s^4} \quad , \tag{5.2}$$

$$\vec{F}_{\mathrm{Tens}} = \beta \frac{\partial^2 \vec{r}}{\partial s^2} \quad . \tag{5.3}$$

Applying this result to elastic light scattering, this model yields correct results in the flexible coil limit [99], but it fails at high stiffness, where it deviates from the solution obtained for rigid rods [101].

In our discrete polymer chain models, the stiffness, i.e. the bending rigidity of the chains composed of N mass points, is introduced into the CG model by the following bending potential

$$U_{\mathrm{bend}} = \frac{\kappa}{2} \sum_{i=1}^{N-1} (\vec{u}_{i+1} - \vec{u}_i)^2 = \kappa \sum_{i=1}^{N-1} (1 - \vec{u}_{i+1} \cdot \vec{u}_i) = \kappa \sum_{i=1}^{N-1} (1 - \cos\theta_{i,i+1}) , \tag{5.4}$$

where $\vec{u}$ is the unit bond vector $\vec{u}_i = \vec{r}_{i+1} - \vec{r}_i / |\,\vec{r}_{i+1} - \vec{r}_i\,|$, connecting consecutive monomers and $\vec{r}_i$ is the position vector to the i-th monomer.

In the numerical model L_p is set by parameter κ of the bonded potential in (5.1). Following Pasquali et al. [178] we use a straightforward definition of the persistence length in the simulation model with respect to the segment length d_0 as

$$L_p = \frac{\kappa d_0}{k_B T} \, . \tag{5.5}$$

In the simulations L_p is varied in the range of $L_p \in \{5, 10, 20, 50\}d_0$, and is thus large compared to one segment length d_0, which is the smallest length scale on which the model chains still exhibit flexibility. This range of L_p is also larger than the persistence lengths used in most previous MD studies of semiflexible chains which were in the range of $L_p \in \{1, ..., 5\}d_0$ [34, 128, 132, 180]. Monodisperse chains with monomer numbers $N = 350$ and $N = 700$ are simulated. Thus, the semiflexible regime of polymer chains for which

$$d_0 \ll L_p \ll L = (N - 1)\,d_0 \tag{5.6}$$

is investigated. Starting from the Langevin Equation of the Rouse model, see [58], the stiffness of the polymer chains can be taken into account by introducing an additional entropic bending force $\vec{F}^B$ of the form

$$\vec{F}^B(s, t) = -L_p k_B T \frac{\partial^4}{\partial n^4} \vec{r}(s, t), \tag{5.7}$$

which can be derived by applying Hamilton's principle to Equation (5.1) [96]. The equation of motion then becomes

$$\xi \frac{\partial}{\partial t} \vec{r}(s, t) = \frac{3k_B T}{2L_P} \frac{\partial^2}{\partial s^2} \vec{r}(s, t) - \frac{3k_B T L_p}{2} \frac{\partial^4}{\partial s^4} \vec{r}(s, t) + \vec{F}^S(s, t) \, , \tag{5.8}$$

where $\vec{F}^S$ is a Gaussian stochastic force. Equation (5.8) is similar to the equation of motion derived by Harris and Hearst [101] and by Harnau et al. [97] who, as in the Rouse model, assume a Gaussian distribution of each segment length. In essence, (5.8) is based on the Rouse model for flexible Gaussian chains and the introduced bending force can be considered as a small and local perturbation of the system. A solution of (5.8) can be achieved in terms of a normal Rouse mode analysis as shown in [208]:

$$\langle \vec{X}_p^{\,2}(0) \rangle = \frac{k_B T}{k_p^{\text{semi}}} = \left[\frac{3\pi^2}{LL_p} p^2 + \frac{3L_p \pi^4}{L^3} p^4 \right]^{-1} , \qquad (5.9)$$

where the effective force constant of semiflexible chains is introduced as:

$$k_p^{\text{semi}} = k_p^b + k_p = (3k_B T L_p \pi^4 p^4)/(L^3) + (3k_B T \pi^2 p^2)/(LL_p) . \qquad (5.10)$$

The value k_p^b is due to the mechanical bending force and k_p arises from the entropic tension. The crossover scaling behavior of single polymers of different persistence lengths L_p is displayed in Fig. 5.3. As one can see, these chains follow the corresponding scaling laws as predicted from theory (5.9). Introducing a critical Rouse mode p_c upon which the crossover scaling from a p^{-2} to p^{-4} behavior, i.e. from entropic tension modes to mechanical bending modes occurs, we can see, that all different simulations, independent of system size N fall on one universal master curve as displayed in Fig. 5.3.

Figure 5.4 exhibits the normal mode amplitudes for melts of linear chains with $N = 70$ and $L_p = 5 d_0$. Also the results for corresponding single chains of the same length are included in Fig. 5.4. Equation (5.9) is plotted in the figure as solid line and the expected p^{-4} scaling is indicated by a dashed line. Good agreement is found for both, single chains and melt results, albeit the theoretical curve slightly underestimates the simulation results at larger p. For comparison and in order to emphasize the enhanced simulations performed in our work, we also show in Fig. 5.3 the incorrect scaling laws published by other researchers in previous studies, which were based on atomistic models.

In some of the previous works on the scaling properties semiflexible chains in a melt an empirical power law $p^{-\alpha}$ with an exponent that

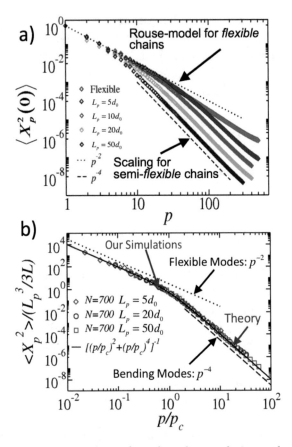

Figure 5.3.: Crossover scaling of single polymer chains and master curve.
a) Transition in the crossover scaling from fully flexible to
semiflexible behavior. b) All different curves lie on one master curve when introducing a dimensionless scaling variable.

ranged from $\alpha = 2.3$ to $\alpha = 3$ was proposed (depending on the
investigator) to describe the dependence of $\langle X_p^2 \rangle$ on p due to the local
stiffness in the chains, based either on simulations [34, 128, 132, 180]
or on theoretical derivations [91]. A p^{-3} power law, as suggested
in [91] and [180], is indicated by the dotted line in Fig. 5.4 a) and
another suggested power law by Bulac et al. [34] is shown in Fig. 5.4 b).

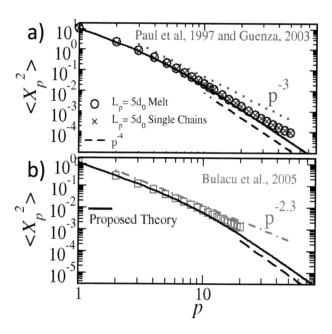

Figure 5.4.: Crossover-scaling of polymer melts and comparison with reported incorrect scaling results of previous, atomistic simulations.

However, from our analysis it can clearly be seen that a p^{-3} and a $p^{-2.3}$ law are insufficient to describe $\langle X_p^2 \rangle$ for larger p values, i.e. larger chain stiffness. Such a power law more or less describes only the behavior in the transition zone around p_c, where no distinct scaling is found and the mean square amplitudes are determined by a combination of Rouse and bending modes.

The reason why the p^{-4} scaling was not really seen in previous simulations is the fact that in most publications only a very limited number of modes ($p < 50$) could be considered because of the small chain lengths ($N \leq 130$). A pronounced appearance of the p^{-4} scaling requires the sampling of a larger range in p-space, where self-similarity with respect to bending modes occurs, that is the persistence lengths must include a sufficient number of monomer segments.

5.2. Coarse-graining of lipid bilayer membranes

In this section we introduce our CG model of lipid molecules which self-assemble in our simulations to form stable bilayer membranes. We analyze the properties of the model at equilibrium and demonstrate that it reproduces key mechanical properties of real lipid bilayers. Non-equilibrium shock wave simulations with our CG model are presented in Chap. 6.

Cell membranes are fascinating supramolecular aggregates that not only form a barrier between different compartments, or organelles, of cells but also harbor many chemical reactions essential to the existence and functioning of a cell [183]. For example, the plasma membrane serves as a barrier to prevent the contents of a cell from escaping and mixing with the surrounding medium. At the same time a cell's plasma membrane must enable the passage of critical nutrients into the cell and the passage of waste products out. Membranes must also be flexible to enable cells to change shape and they must have malleable topologies such that a cell can grow and divide into two separate parts, each of which has a completely closed contiguous membrane, a vesicle. Membranes in living biological systems have managed to balance these multiple demands by exploiting the special amphiphilic physical properties of the molecules that make them up.

The most common membrane constituents are lipid molecules with two physically separated subdomains, usually an elongated hydrophobic domain made up of fatty acid tails, associated with a hydrophilic head group. The most abundant lipids in cell membranes are the phospholipids, molecules in which the hydrophilic head is linked to the rest of the lipid through a phosphate group [6], cf. Fig. 5.5. The hydrophilic head groups of lipids dissolve readily in water because they contain polar groups that are easily incorporated in the hydrogen-bonding network of the surrounding water. In contrast, the hydrophobic hydrocarbon tails are uncharged and non-polar and thus try to aggregate in energetically and entropically favorable structures that minimize their contact with surrounding water molecules. The nearly cylindrical shape of most membrane lipids makes the bilayer the

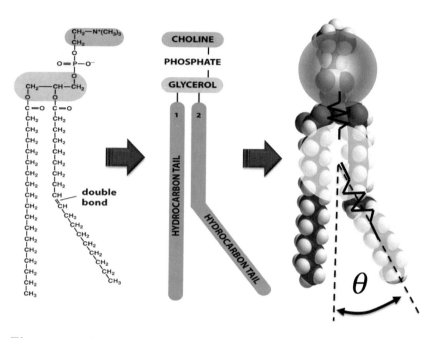

Figure 5.5.: An example of a common phospholipid (dipalmitoylphosphatidylcholine, DPPC, chemical formula $C_{40}H_{80}NO_8P$) and its coarse grained representation. The CG model is composed of three parts, one hydrophilic head (H) bead and two hydrophobic tail (T) beads, connected by harmonic bonds. Angle θ is the angle of the bending potential of (5.4). We note here, that our model is by no means restricted in the number of head or tail particles: We just decided here to use the simplest possible representation of our model which involves only three particles [196].

most common geometrical organization for spontaneous self-assembly of lipid molecules in aqueous environment.

The fluid mosaic model of membranes by Singer and Nicolson [200] was of great heuristic value, and contributed to our general thinking about membrane organization by providing insights into the assembly of lipids into membranes resulting from diffusion processes. However, subsequent experimental evidence indicated that the lateral motion of membrane components was not free after all, but constraint by various molecular mechanisms, such as direct or indirect interactions with elements. Many questions pertaining to cell membrane dynamics remain unanswered until today and many processes such as lateral segregation or transversal asymmetry are known to occur in membranes on vastly different spatiotemporal scales [205, 230]. For example, membrane vesicle fusion in vivo has been reported to occur on a length scale of tens of nanometers and a timescale that is sub-millisecond, possible faster than 100 microseconds [146, 156]. It is therefore not yet possible to visualize the re-formation of the lipids in a fusion pore experimentally. It is precisely at the spatial and temporal limits where instruments for direct observation fail that the membrane concept becomes fuzzy and unclear.

On the other hand, all-atom MD simulations of lipid bilayers which resolve the dynamics of individual atoms are limited to fairly small membrane samples (tens of nanometers in extension) and very short time scales of at most a few hundred nanoseconds, [27] leaving CG simulations as the only currently available computational tool to access mesoscale phenomena such as the dynamics of molecular self-assembly of lipid molecules.

There is a very large body of literature on computational studies of the static and dynamic properties of biomembranes using atomistic and CG modeling approaches, see e.g. [40, 108] which have been reviewed in depth e.g. by Pandit and Scott [177] and Woods and Mulholland [246].

With the rise of so-called *solvent-free*, or *implicit* simulation schemes for membranes, which became fashionable at the turn of the millennium, the number of publications in this field has constantly increased. Solvent-free models of lipid bilayer structures do not explicitly take the fluid molecules of the aqueous environment into account. These models are either based on a Langevin equation of motion accounting for the Brownian random motion of the fluid molecules, or on the complete modeling of all effects of the solvent by effective interaction potentials between the constituent particles of the membrane only [28, 75, 172, 254]. As one is usually only interested in the structural and dynamic details of the membrane and not in the surrounding fluid, this allows to reduce the number of necessary integrations and thus the computational costs considerably. Many of the used potentials are either derived from intuition, from standard potentials routinely being using in polymer physics or from quantum chemical calculations of force field parameterizations [40, 179, 202, 243].

5.2.1. Lipid-lipid and lipid-water interactions

Lipid bilayer membranes in biological cells are complex structures composed mainly of phospholipids, and, to a lesser extent, of membrane proteins and various carbohydrates. Here, we do not attempt to model all these complex details. Instead, we employ a simple CG model of a complex phospholipid, which forms stable bilayers that reproduce key mechanical parameters of cell membranes, such as bending rigidity R_B, area compression modulus K_a, rupture strength S_r, and mass density profile. In our work we introduce a three-particle model of a DPPC molecule according to the scheme in Fig. 5.5. Hence we have three interaction sites: one head and two tail beads.

This model has the advantage that it can be easily adjusted to yield different membrane rupture strengths and bending rigidities, thus allowing a systematic study of how shock wave induced damage is affected by these mechanical parameters [82]. The structural and amphiphilic properties of the lipids are modeled by interaction potentials according to Equations (5.11)–(5.14). The mass points of a lipid chain are connected by anharmonic bonds with the potential

$$\phi_{\text{bond}}(r) = \begin{cases} -\frac{1}{2}\kappa\, r_\infty^2 \ln\left[1 - \left(\frac{r}{r_\infty}\right)^2\right] & \text{for} \quad r < r_\infty\,, \\ \infty & \text{for} \quad r \geq r_\infty\,, \end{cases} \quad (5.11)$$

where κ is the force constant and r_∞ denotes the maximum bond length.

The finite size of a monomer of the lipid molecule is taken into account by the truncated Lennard-Jones (LJ) potential

$$
\phi_{\mathrm{LJ}}(r) = \begin{cases} 4\epsilon \left[\left(\frac{\sigma}{r}\right)^{12} - \left(\frac{\sigma}{r}\right)^{6} \right] + \epsilon & \text{for} \quad r < r_{\mathrm{cut}}, \\ \infty & \text{for} \quad r \geq r_{\mathrm{cut}}, \end{cases} \tag{5.12}
$$

where r denotes the distance between two (non-bonded) mass points. In our simulations we set the bond length $r_\infty = \sigma$ and the interaction cutoff distance $r_{\mathrm{cut}} = \sqrt[6]{2}$.

To straighten the lipids, a bending potential for all bonds depending on the bond angle θ (see Fig. 5.5) is used in the form

$$
\phi_{\mathrm{bend}}(\theta) = \lambda \left(1 - \cos\theta\right), \tag{5.13}
$$

where λ determines the stiffness of the lipid molecule.

The typical structures of self-assembling lipids arise from the hydrophobic and hydrophilic interactions of the lipid tails and heads, respectively, with the surrounding water molecules. When using a solvent-free model, the hydrophobic interactions of lipid tails with water can be modeled using an attractive potential between all tail particles of different lipid chains as indicated in Fig. 5.5:

$$
\phi_{\mathrm{attr}}(r) = \begin{cases} -\alpha & \text{for} \quad r < r_{\mathrm{cut}}, \\ -\alpha \cos^2 \left(\frac{\pi(r - r_c)}{2h_c} \right) & \text{for} \quad r_c \leq r \leq r_c + h_c, \\ 0 & \text{else}, \end{cases} \tag{5.14}
$$

where α is a constant and h_c determines the range of the attractive potential, i.e. the larger h_c the greater the hydrophobicity of the lipid tails, and r_{cut} is the same cutoff as in (5.12). A tunable potential of this type (involving a cosine function) was introduced by Steinhauser [206] into polymer physics for the realistic MD simulation of polymer-solvent interactions with varying solvent qualities. A similar form of this potential which we use in (5.14) has then later been adopted to be used in implicit solvent MD simulations of lipids by Cooke et al [46].

The decay range h_c of (5.14) has direct influence on bending rigidity and rupture strength of the membrane, which can be conveniently exploited to simulate lipid bilayers with a broad range of mechanical properties. While this model was originally intended to describe stable fluid bilayer without the need for additional solvent, i.e. to save computation time, we explicitly model the aqueous environment here to solvate our membrane structure in a way that can be used later for shock wave simulations. Since here we are interested only in characterizing equilibrium properties of our membrane model, we don't need to use our new multiscale coupling scheme introduced in Chap. 3. Therefore, we model water with a CG model by Chiu et al. [43], which in essence is based on a simple, short-ranged Morse potential:

$$\phi_{H_2O}(r) = \begin{cases} \gamma\varepsilon\left[e^{\beta\left(1-\frac{r}{\sigma}\right)} - 2e^{\frac{1}{2}\beta\left(1-\frac{r}{\sigma}\right)}\right] & r \leq r_{c,H_2O} \\ 0 & r > r_{c,H_2O}, \end{cases} \qquad (5.15)$$

with fit parameters $\beta = 7$, $\gamma = 1.31911$, and interaction cutoff $r_{c,H_2O} = 2.544\,\sigma$. One water particle in the model represents four H_2O molecules. The CG potential of (5.15) reproduces the heat of vaporization, surface tension and compressibility of water with good accuracy. The interaction of membrane head particles with water beads is also given by (5.15), while the hydrophobic lipid tail particles repel water particles according to (5.12). If wanted, a mapping from dimensionless simulation units to absolute units can be done by specifying $\varepsilon = 310k_B T$ and $\sigma = 0.629$nm. Particle masses are $m_{H_2O} = 72$gmol^{-1}, $m_{tail} = 115$gmol^{-1}, and $m_{head} = 220$gmol^{-1}. As we proceed to show in the next section, this choice of parameters results in a mass density profile of the solvated lipid bilayer that is in reasonable agreement with the calculated mass density in all-atom simulations, cf. Fig. 5.6.

5.2.2. Distribution of the mass density

During shock wave impact (which we discuss in Chap 6), the lipid bilayer is subjected to large accelerations. Hence, it is important for our CG model to have a realistic mass distribution in order to correctly account for inertial effects. Accordingly, in Fig. 5.6 we compare the

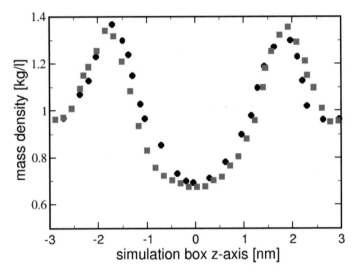

Figure 5.6.: Mass density distribution of an atomistic DPPC membrane model (dark circles), measured along the direction normal to the bilayer planes and our generic CG model (light squares). Both simulations were done with the author's software suite `MD-CUBE` [206].

mass density distribution of the employed CG model, measured along the direction normal to the bilayer plane, with the results of an *all-atom* simulation of DPPC. The agreement is good with the density peaks at the hydrophilic head groups and the depletion zone at the bilayer center being well reproduced.

5.2.3. Phase diagram of our bilayer membrane model

The attractive range of the lipid tail interactions is governed by the decay range h_c, as discussed above. We find stable bilayer structures in the interval $h_c = [0.8, 1.4]\,\sigma$ at a temperature $T = 310\,K$, which corresponds to physiological conditions. For $h_c \lesssim 0.7\sigma$, the bilayer dissolves, while it is crystalline for $h_c \gtrsim 1.5\sigma$. In Fig. 5.7 we display the phase diagram of our lipid membranes based on the two parameters temperature T and attractive potential range h_c. Each data point was generated from averaging and analyzing 20 different simulations for the specific combination of (T, h_c). The fluid character of the bilayer within the stable range is verified by computing the in-plane diffusion coefficient $D = 0.03 - 0.06\sigma^2/\tau$ and observing a non-zero flip-flop rate, i.e., the probability per unit time that a single lipid changes from one surface to the other. The phase diagram of the membrane structures of our model was calculated by averaging 100 simulation runs for each displayed combination of temperature T and attractive potential range h_c.

5.2.4. Order parameter

In order to characterize the degree of structural order in the system of lipid molecules during self–assembly (cf. the simulation snapshots in Fig. 5.7), we introduce an order parameter $O_2(\tau)$ based on the second order Legendre polynomial $P_2(\cos\vartheta)$:

$$O_2(\tau) = \left\langle P_2\big(\cos\vartheta(\tau)\big)\right\rangle = \left\langle \frac{3}{2}\cos^2\vartheta(\tau) - \frac{1}{2}\right\rangle . \tag{5.16}$$

Here, $\vartheta(\tau)$ denotes the momentary angle between the lipid direction vector and the average normal vector $\vec{n}(\tau)$ of the membrane plane, called director, at time τ. The lipid direction vector is defined as the vector connecting the center of mass of a head particle with the center of mass of the second tail particle. The brackets $\langle...\rangle$ indicate averaging over all lipids. $O_2(\tau)$ can be averaged over many time steps to obtain one single average value, but, as we show in Fig. 5.8, it is more interesting to look at the information provided by plotting the development of $O_2(\tau)$ with simulation time. $O_2(\tau)$ provides information about the anisotropy of a system, i.e. the directional orientation of our self–assembled membranes. The values of $O_2(\tau)$ are in the range of $S_2(\tau) \in [-0.5, 1.0]$. For $O_2(\tau) = -0.5$, the molecules exhibit an alignment similar to a bottle brush, and hence, the average normal vector equals 0. $O_2(\tau) = 0.0$ occurs, when the lipids are isotropically oriented at random, like in a non-directional fluid phase. Finally, $O_2(\tau) = 1.0$

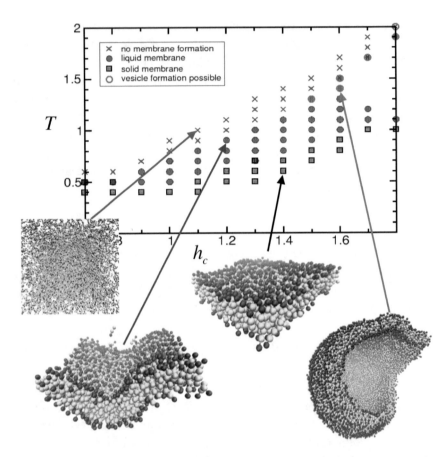

Figure 5.7.: Temperature–attraction (T, h_c) phase diagram of our lipid bilayer membrane model with some typical representations of membrane structures [221]. We can clearly distinguish several regions where the lipids do not form stable structures (red), or fluid structures (blue), crystalline-like structures (black) with a minimum amount of fluctuations and finally a stable region where the lipids form stable vesicles (completely closed bilayer membranes which can serve as a model of the plasma membrane of eukariotic cells.

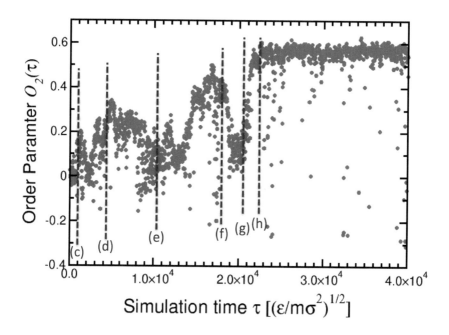

Figure 5.8.: Order parameter $O_2(\tau)$ vs. simulation time τ during self-assembly of lipids in the stable bilayer regime. The indicated letters (c)-(h) in this figure correspond to certain points in time during the self-assembly: (c) $\tau = 1,000$, (d) $\tau = 4,000$, (e) $\tau = 10,200$, (f) $\tau = 18,000$, (g) $\tau = 20,200$ and (h) $\tau = 22,400$. For better presentation of data points, (a) and (b), which are very close to $\tau = 0$ are not labeled.

corresponds to a perfect parallel alignment of the lipids. Typical liquid crystals exhibit values between $O_2(\tau) = 0.3$ and $O_2(\tau) = 0.8$. In Fig. 5.8 we present a plot of $O_2(\tau)$ versus simulation time of a typical system that finally forms a stable liquid bilayer membrane, cf. Fig. 5.7. Starting from a completely disordered, random initial state with $O_2(0) = 0.0$, the system slowly develops into an equilibrium state with values of $O_2(\tau)$ saturating just below $S_2 \approx 0.6$ which is within the typical range of liquid crystals for a fluid phase. Thus, our model membranes also exhibit fluid-like behavior. The undulations

of the values of $O_2(\tau)$ in Fig.5.8 are due to this fluidity property of
the simulated membranes. After simulation time $\tau = \tau_{eq} \approx 22400$
(snapshot (h) in Fig. 5.8), the system obviously has reached its equilibrium state, as indicated by the fluctuations of the order parameter
about a mean value. There are still strong fluctuations of $O_2(\tau)$ for
$\tau > \tau_{eq}$, indicating that $O_2(\tau)$ is a very sensitive quantity for local
rearrangements of molecules, even in a stationary state. Whenever a
rearrangement of lipids due to membrane fusion occurs, $O_2(\tau)$ drops
while the lipids rearrange, and then rises again, indicating increased
order in the newly formed bilayer. Hence, the fluctuations of $O_2(\tau)$
can be used as an indicator for major molecular rearrangements during
the self–aggregation of lipids into bilayer membranes.

In Fig. 5.9 we present two striking examples of systems with $1,000$
lipids for which we performed a detailed analysis of the order parameter $O_2(\tau)$. In Fig. 5.9 (a) we show $O_2(\tau)$ for a system at very low
temperature $T = 0.5\varepsilon/k_B$ for a variety of ranges of the attractive
potential h_c of the lipid tails which indicates the degree of hydrophobicity. The aggregated systems at such a low temperature are
in a crystal-like solid phase, with a very particular arrangement of
molecules. Interestingly, the fluctuations of $O_2(\tau)$ are the largest for a
small range of hydrophobic interactions, tailing off for larger h_c-values.
$O_2(\tau)$ is generally limited to a rather narrow range of values within
the interval $[0.5, 0.7]$ and some fluctuations going down to $O_2(\tau) = 0.4$.
Hence, we see that reducing the temperature to a rather low value has
a dramatic effect on the overall structures of the aggregates that are
formed. However, we also realize that tuning the interaction range h_c
in our model has an effect on the resulting bilayer structures that is
much less pronounced than the effect of varying temperature T and
keeping h_c at a constant value, which is shown in Fig. 5.9 (b).

Here, we show the results for $O_2(\tau)$ for a system with $h_c = 1.3$,
varying the temperature between $T = 0.5$ and $T = 1.3$. Hence, the
smallest value of T in Fig. 5.9 (b) represents the same conditions as in
Fig. 5.9 (a) for the largest h_c-value. We see, that for these two cases,
$O_2(\tau) = 0.7$ consistently in (a) and (b) with very small fluctuations.
It should be noted, that in (b) the range of $O_2(\tau)$ is much larger than

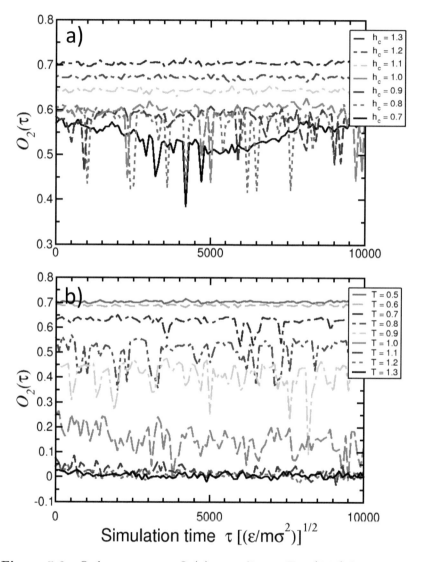

Figure 5.9.: Order parameter $O_2(\tau)$ according to Eq. (5.16) for a system
at equilibrium with $N_{lip} = 1,000$ [221]. a) A system at temperature $T = 0.5\varepsilon/k_B$ with different ranges of hydrophobic
tail interactions, indicated by parameter h_c. b) A system
with $h_c = 1.3\sigma$ for a range of different temperatures T. Note
the different scales for the y-coordinate in a) and b).

in (a), ranging from random configurations with $O_2(\tau) \approx 0$ for the largest temperatures, where no clustering of lipids occurs, up to values pertaining to strongly aggregated bilayer membrane systems, which is confirmed by visual inspection of the self–assembled structures. The intermediate range of temperatures in Fig. 5.9 (b) pertain to stable bilayer membrane structures in a fluid phase which lead to distinct fluctuations of the order parameter $O_2(\tau)$. Hence, from Fig. 5.9 we can induce that the degree of order obtained in the self–assembled lipid aggregates for our model is much more sensitive to the variation of temperature rather than the variation of interaction range h_c.

5.2.5. Pair correlation function

Another quantity that provides information about the local molecular equilibrium structure of the aggregated lipids is the pair correlation function

$$g(r) = \frac{V}{4\pi r^2 N^2} \left\langle \sum_i \sum_{j \neq i} \delta(r - r_{ij}) \right\rangle. \qquad (5.17)$$

In Fig. 5.10 we analyze $g(r)$ for the same systems discussed in Fig. 5.9, where, in the case of (a), the temperature is set to a very low value $T = 0.5\varepsilon/k_B$, and in the case of (b), h_c is set to a constant value of $h_c = 1.3\sigma$. In Fig. 5.10 (a) $g(r)$ exhibits pronounced peaks for all values of the hydrophobicity parameter which is varied in the range $h_c \in [0.7, 1.3]$. These peaks indicate a very strong overall aggregation of the lipids and are all the more pronounced with increasing scope of the hydrophobic interaction. The peaks are clearly visible at least until a distance of $r = 6\sigma$. Thus, the structure of these systems is that of a crystal-like solid with a pronounced long-range order in accordance with our conclusions drawn for the order parameter $O_2(\tau)$ in Fig. 5.9 (a). In Fig. 5.10 (b) where the temperature is varied, we can see that for $g(r)$ the resulting aggregated membrane structures at equilibrium are more sensitive to a change in temperature T than to a change of the hydrophobic interaction range. For the lowest

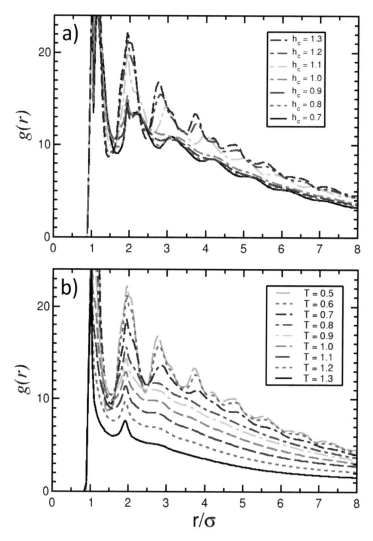

Figure 5.10.: Pair correlation function $g(r)$ according to (5.17) for a system of $N_{lip} = 1000$ lipids at equilibrium [221]. The analysis is done for the same systems discussed in Fig. 5.9. a) displays a system with very low temperature $T = 0.5\varepsilon/k_B$ for different values of hydrophobicity h_c. b) shows a system with $h_c = 1.3$ for different temperatures T in units of ε/k_B.

temperatures the system is practically frozen into a quasi-crystalline state, exhibiting long-range order even at distances beyond $r = 6\sigma$. With increasing temperature the system attains a different structure with only local order up to distances of roughly $r = 3\sigma$. Hence, in this range of temperature, the structural features of the aggregated lipids are similar to fluid behavior with fluctuations in the bilayer membrane height and shape. Finally, for very large temperatures, only the first and second nearest neighbor peaks remain which reflects in essence the local structure of single lipid molecules. This means that no aggregation of lipids occurs for this range of temperatures and the given value of h_c, and no bilayer membranes are formed. All of these results are consistent with our discussion of the order parameter $O_2(\tau)$ in Fig. 5.9 (b) and also with the phase diagram of our model presented in Fig. 5.7.

5.2.6. Elastic modulus

The bending modulus M_B determines by how much an applied external force deforms a lipid bilayer from its perfectly flat energetic ground state. It can be obtained from linear response theory by analyzing the bilayer height fluctuation spectrum. Helfrich's linearized continuum theory [199] predicts the following relation for a 2D elastic sheet:

$$S(n) = \langle c_{\vec{n}} \, c_{\vec{n}}^* \rangle = \frac{k_B T L_{\parallel}^2}{16\pi^4 (M_B \vec{n}^4 + s\vec{n}^2)}, \qquad (5.18)$$

where

$$c_{\vec{n}} = \frac{1}{N} \sum_{j=1}^{N} (x_j - x_0) \exp\left[\frac{-2\pi i}{L_{Box}} (n_y y_j + n_z z_j) \right] \qquad (5.19)$$

is a Fourier component of the deviation of the local bilayer height calculated with respect to its average height x_0. The sum runs over all bilayer particles with coordinates x_j, y_j, and z_j. Vector $\vec{n}$ is a 2D vector with components n_y and n_z in reciprocal space, which relates to a wave vector through $\vec{q} = 2\pi\vec{n}/L_{Box}$. The value L_{Box} is the edge

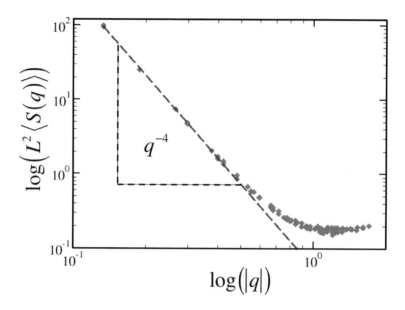

Figure 5.11.: Membrane height fluctuation spectrum for a bilayer system with $N_{lip} = 10^4$ lipids, interacting with attraction parameter $h_c = 1.0\sigma$. Scaling of the spectrum follows the relation $S(q) \propto q^{-4}$, as predicted by continuum theory.

length of the simulation box in x-direction which is the direction of shock wave propagation. Expressing (5.18) in terms of wave vectors, we obtain the scaling law

$$S(q) \propto q^{-4} \,. \tag{5.20}$$

However, (5.20) is only valid in the regime below the crossover wave vector $q = \sqrt{\sigma/M_B}$, as $S(q)$ probes the internal structure of the lipid bilayer at higher wave vectors. As shown in Fig. 5.11, the asymptotic q^{-4} scaling regime is recovered at this system size, which implies that the characteristic lipid bilayer bending modes can be accommodated in the simulation box. We obtain the bending rigidities by fitting (5.18) to the simulation data.

Table 5.1.: Calculated bending stiffness M_B and compression modulus K_A for bilayers with different lipid attraction parameters h_c and simulation box size $L_{Box} = 80$nm.

h_c/σ	$M_B/k_B T$	$K_A/10^{-3}\text{Nm}^{-1}$
0.8	13.7 ± 0.4	178 ± 7
1.0	14.4 ± 0.9	196 ± 4
1.2	16.2 ± 0.4	222 ± 6
1.4	21.3 ± 0.5	235 ± 4

Continuum theory relates M_B to the area compression modulus via $K_A = 24 M_B/d^2$, with d being the bilayer thickness, which we take to be 5nm [175]. Results for these quantities are provided in Tab. 5.1. We note that the values obtained from our simulations are well within the experimental range of biomembranes [139, 175], which is $K_A \approx 237 \times 10^{-3}\text{Nm}^{-1}$, and $M_B = (10 - 50)k_B T$, depending on the technique used for optical measurements.

6. Laser-induced shock wave destruction of human tumor cells: experiments and simulations

In this section we present a combined experimental/numerical approach to investigate the destructive effects of shock waves in biological soft matter systems. On the one hand we present our wet lab shock wave experiments with human brain tumor cells, U87 glioblastoma, which we expose to very well-defined shock waves generated by laser ablation.

On the other hand – in simulations – we focus on one major structural component of cells that is responsible for their mechanical stability: the plasma membrane. Here, we present shock wave simulations of large membranes performed with our new multiscale coupling scheme discussed in Sect. 3.6, with which we are able to couple a coarse-grained, particle-based membrane model with the surrounding fluid which is modeled as a continuum. We introduce an order parameter to measure the degree of damage induced in a membrane and show how this parameter depends on the shock front velocity. We observe self-repair in the membrane after shock wave impact which has also been reported in experiments. We further find the existence of a limiting shock wave speed beyond which damage in a membrane becomes irreversible.

In the shock wave experiments with cells we managed to improve considerably existing experimental approaches in the life sciences to investigating shock wave effects in eukariotic cells. Furthermore, we determine the pressure threshold necessary to achieve a convincing

© Springer Fachmedien Wiesbaden GmbH 2018
M. O. Steinhauser, *Multiscale Modeling and Simulation of Shock Wave-Induced Failure in Materials Science*,
https://doi.org/10.1007/978-3-658-21134-9_6

biological destructive effect in the tumor cells which we examine with several independent methods: cell viability measurements using trypan blue, cell counting with a Coulter counter and a MTT viability test over a period of 5 days and several cell cycles using appropriate controls. Our findings are that shock waves generated by laser ablation are able to completely destroy U87 tumor cells if the shock front reaches a critical pressure level beyond roughly 80 MPa.

Exploring the potential of shock waves for destroying or damaging cancer cells possibly can open a new road for medical tumor treatment, avoiding the disadvantages of currently established methods, by exploiting only the *mechanical* destructive effects of a shock wave interacting with cells. Besides resection, i.e. the surgical removal of tumors, and chemotherapy, an established treatment in tumor therapy is the use of **HI**ghly **F**ocused **U**ltrasound (HIFU). The treatment with ultrasound is based in essence on the destruction of tissue by heating. At temperatures above 56 degrees Celsius proteins start to coagulate, causing the cells to go into apoptosis, i.e. programmed cell death. Additionally, HIFU leads to the generation of small bubbles in the tissue, growing fast and eventually imploding, which causes shock waves expanding through the tissue on a length scale of micrometers. This phenomenon, known as cavitation, leads to additional cell and tissue damage both, in the cancerous and healthy tissue. Principal problems in focusing of ultrasound usually lead to unwanted side effects such as cutaneous burns, and also the HIFU sessions have to be repeated several times. In order to better monitor the treatment, sessions are often done in a magnetic resonance tomograph (MR-Guided HIFU) which renders this form of therapy very expensive.

On the other hand, chemotherapy for tumor treatment also has many unwanted side effects and surgery principally involves a high risk for the patient. By way of using the mechanical effects of shock waves rather than relying on heating of tissue it seems possible to destroy or damage the cytoskeleton or the plasma membrane of cells, as shock waves carry more energy in a shorter time interval than ultrasound waves do. The complexity occurring in the interaction of shock waves with soft, biological matter requires a combined approach

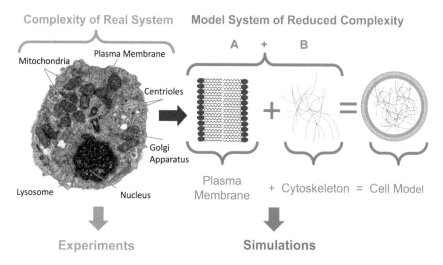

Figure 6.1.: We adopt an integrated approach combining the full complexity of the real experimental system with computational
CG models of reduced complexity. In CG models, typically
only two major components of cells determining their mechanical properties relevant for their interaction with shock
waves are considered: The plasma membrane and/or the
cytoskeleton. Figure adapted from [223].

using experimental and numerical methods which is exemplified in
Fig. 6.1.

6.1. The impact of shock waves on tumor cells

When irradiating an absorbing material with a pulsed laser, the optical
energy deposited on the absorber is transformed into mechanical energy.
A shock wave forms at the surface and then propagates through the
absorber [35]. The shock wave properties such as rise time, velocity of
propagation or peak pressure depend on the absorbing material and
the laser parameters. For one specific laser/absorber system, the peak
stress of the shock wave can be tuned by varying the laser fluence that

is equal to the total energy deposited per area of illumination [232]. In this experimental configuration, well-defined, reproducible shock waves can be generated without the side effect of heating or cavitation [59]. Thus, the pure mechanical effect of the shock wave on the cells can be investigated.

6.1.1. Preliminary tests of experimental setups for laser-induced shock wave generation

In many early studies of the 1980s and 1990s, the effects caused by shock waves in cell cultures could not always be clearly separated from those caused by direct laser irradiation or probable contamination of cell cultures which were used as *in vitro* models. In addition, a precise characterization of the shock wave profile was very often not done due to a lack of precision of hydrophone measurements, and proper control experiments were not always performed or reported. The repetitive use of the same polymer – usually polyimide (PI) or polystyrene (PS) – as target material for shock wave generation, can – as we know today – diversify the resulting profile of shock waves due to changing absorbance of the polymer even after the first laser pulse [253].

Figure 6.2 shows a high-speed photography of one of our preliminary laser ablation experiments where we reproduced the experimental setup of several very early studies performed by Doukas and others [59, 61, 122, 167, 140, 141] which used standard cuvettes made of PS or PI for shock wave generation. Here, the laser beam is aimed directly onto the bottom of a cell culture vessel. Upon treatment with the laser, a shock wave is formed that propagates through the bottom into the vessel and interacts with the cells. For beam diameters of a few millimeters, peak pressures just beyond ~ 30 MPa can be obtained. Despite the relatively low peak pressures in these experiments, in several of the early studies damage and permeabilization of mammalian cells were reported [59, 60, 61, 62, 122].

As can bee seen in the photographs of Fig. 6.2, in the experimental setup of early studies the PI or PS films often formed one end of

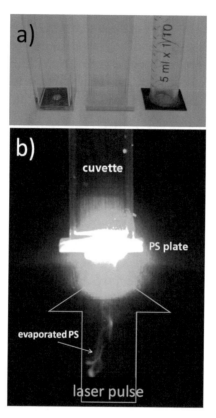

Figure 6.2.: Replicated setup of early experimental studies of shock
wave interactions with biological cells, where the shocks are
generated by laser-ablation. a) From left to right: cuvette
sealed with a black PS plate (0.5mm thick) using method 2;
cuvette sealed with a white PS plate (2.0mm thick) using
method 1; pipette (5ml, cut into a 5cm piece) and sealed
with a black PS plate (0.5mm thick) using method 1. b)
High-speed photograph of a cuvette sealed with a white
PS plate at the bottom, exposed to a single laser pulse
of 532nm wavelength. The apparent laser radiation seen
in the interior of the cuvette is just light scattered off the
cuvette walls. Below the PS plate, a PS vapor cloud can be
observed. Methods 1 and 2 are explained in the main text.
Photographs are © Martin O. Steinhauser [212].

a narrow pipette tube in direct contact with the target cells, which decreased the reproducibility of the experiments considerably [61, 122], as we show in our preliminary tests displayed in Figs. 6.3–6.6. Thus, studies of the biological effects of shock waves induced by pulsed laser ablation of polymer films have generally been hindered by difficulties in reproducible growth conditions of cell cultures and by the difficulty of generating well-characterized and reproducible pressure pulses whose temporal and spatial characteristics are known. For future applications in the medical sciences it is essential to study not only single cells but more complex systems such as tissues. This is one of the challenges for future studies in the field.

In one of the early studies of shock wave effects on cells it was concluded that for one specific cell line the survival rates of cells exposed to laser-induced shock waves depend on the stress gradient $\sigma = p_{max}/\tau_r$, where p_{max} is the peak pressure and τ_r is the rise time of the shock wave [61]. The rise times of the shock waves in this study were varied between 10ns and 30ns. However, repetitive laser shots (5 times) were used which renders the results of these experiments less reproducible and reliable. In fact, the survival rates among different cell lines differ remarkably at constant physical parameters (p_{max}, τ_r). For example, only 50% of transformed (immortalized) retinal pigment epithelium cells survive exposure to shock waves with $p_{max} = 74$MPa and $\tau_r = 10$ns. However, 100% of normal retinal pigment epithelium cells survive this procedure [59]. A shock wave with $\tau_r = 10$ns and $p_{max} = 30$MPa kills 50% of mouse breast sarcoma cells [61], whereas human promelocytic leukemia cells survive this exposure [122].

For generating shock waves in fluid medium in a test tube, transparent PS cuvettes with a volume of 5 ml turned out to be a good choice, cf. Fig. 6.2. Two vessels were prepared for the ablation experiment following the documentation of previous experiments in the literature which report two different methods of preparation, cf. Fig. 6.2 a):

- **Method 1:** The bottom of the test tube is cut off and substituted by a PS plate using superglue.

- **Method 2:** The PS plate is directly glued to the container bottom, again using superglue.

Note that the PS plate just acts as a medium that absorbs the laser energy and transforms it into a mechanical shock wave propagating through the bottom of the vessel into the fluid. Hence, the laser beam just hits the PS plate glued to the bottom of the vessel – the interior of the test tube is not exposed to the laser radiation. According to [122, 167] the shock waves are generated either by laser ablation or instant thermal extension of the target or a combination of both, and launched into the aqueous medium within the vessel.

After laser beam exposure, we observe a dramatic change of the surface structure of the polystyrene plate, see Fig. 6.3. Here, the PS plate was exposed to multiple laser pulses. The micrographs a)-d) in Fig. 6.3 exhibit the section (darker region) that was exposed to a *single* laser pulse in a) and up to 5 laser pulses in d). The number of laser pulses applied to the surface is displayed in each micrograph. Figure 6.3 a) clearly shows an inhomogeneous laser beam profile. Because of the non-uniform distribution of the laser beam intensity, some regions are more coagulated than others. This is also verified in the microscopic 3D analysis in Fig. 6.3 e). When several laser pulses are applied to the same plate position, a black halo is formed around the impact region. The bright spots within the black regions in Fig. 6.3 a)-d) are the ash-remains of coagulated material.

In Fig. 6.4 a) and b) 2D and 3D microscopic images of a PS surface are shown. The surface was exposed to a nanosecond laser pulse of a Quantel laser at 532 nm wavelength. The laser beam was focused with a 100mm lens which resulted in a focused diameter of $500 \mu m$. Fig. 6.4c) demonstrates that directly exposing the PS bottom of a cuvette leads to a non-linear and inhomogeneous absorption of laser energy. As a result, which is also shown in Fig. 6.5, the generated shock waves are not replicable if laser pulses are applied several times.

Finally, in Fig. 6.6 the shock wave profiles of the experiments shown in Fig. 6.4 are displayed. The different colors correspond to the number of laser pulses applied, where the black line corresponds to the pressure

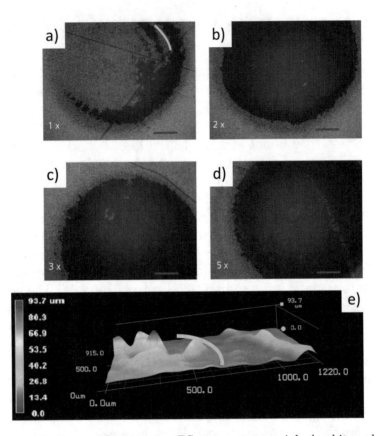

Figure 6.3.: Laser ablation using PS as target material. A white polystyrene plate (2mm thickness) is exposed to: A single laser
pulse in a), two laser pulses in b), three laser pulses in c),
and five laser pulses in d). The diameter of the black area is
roughly 6mm. The yellow line in micrograph a) corresponds
to the 3D microscopic height profile displayed in e). The
red scale bars in the micrographs exhibit a length of 1mm.
e) Microscope image of the 3D height profile of a surface
section after one laser pulse. The yellow line corresponds to
the yellow line in micrograph in a). Microscopic images are
© Martin O. Steinhauser.

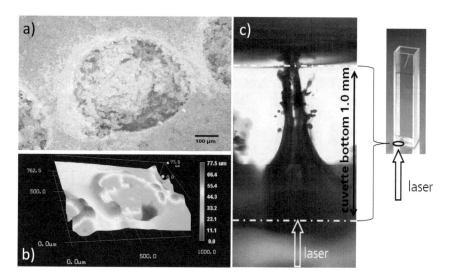

Figure 6.4.: a) A white PS plate was exposed to a focused laser beam. The diameter of the focus was $\approx 500\,\mu m$. b) 3D microscopic profile of the micrograph in a). c) The transparent bottom of a PS cuvette was *directly* exposed to a 580 mJ pulse of a HIGH Q picosecond laser. The exposure time was about 2 seconds at maximum energy at $\lambda = 1,064$nm. The focus diameter was $50\mu m$. The white dashed–dotted lines indicate the cuvette bottom with thickness 1mm. Photographs are © Martin O. Steinhauser.

profile of the first laser pulse which is just below 1MPa and thus way too small in oder to generate any damage to cells. It can also be seen, that the results of shock wave generation with this setup are not at all reproducible.

6.1.2. Hydrophone specification

Shock waves in a fluid are measured using a calibrated polyvinylidene fluoride (PVDF) needle hydrophone with a 0.5 mm diameter sensitive element, see Fig. 6.7. The resolution of the rise time of the hydrophone is roughly 25ns. The calibration factor of a hydrophone which is

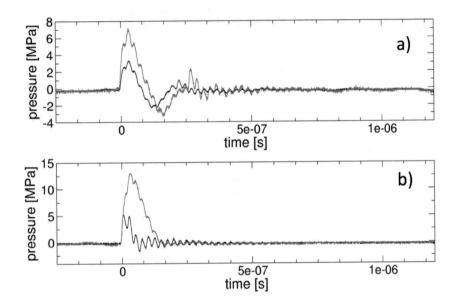

Figure 6.5.: Pressure levels measured with two cuvettes exposed to different energies: a) 180mJ pulse energy and the hydrophone
was placed at a distance of 1.1cm. b) 580mJ pulse energy
and the hydrophone was placed at a distance of 0.65cm.
The two curves in each subfigure display from top to bottom the first and second laser pulses. One again recognizes
the large difference in pressure when exposing the same
cuvette to laser pulses multiple times.

needed for calculating the pressure in absolute units from the potential
difference measurements is determined as follows:

$$U_{\text{probe}} = \frac{Q_{\text{probe}}}{C_{\text{sum}}}, \qquad (6.1)$$

where U_{probe} is the sensitivity of the probe in Volt/bar, Q_{probe} is
the sensitivity of the probe in pC/bar, and C_{sum} is the sum of probe
capacity including the cable (C_{probe}) plus the capacity of the connected
device (amplifier or recorder input capacity; quite often 13pF or 50pF)

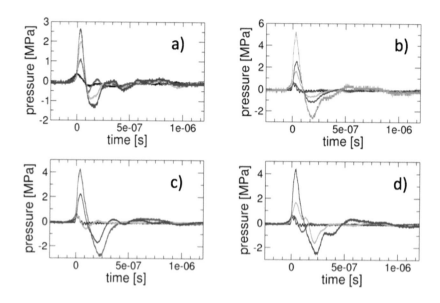

Figure 6.6.: Experimental pressure profiles of four different white 2.0mm
PS samples corresponding to the four experiments a)-
d)displayed in Fig. 6.3. Colors correspond to the number
of pulse repetitions. black: 1, red: 2, green: 3, blue: 5. Note
the different heights of the peak pressures in the various
samples.

plus, if necessary, the capacity of an extension cable in pF. For the
presented experiments in this work, the capacity of the connected
device is 50pF. Two different hydrophones were used in our experiments.
The specific calibration factors of the used hydrophones are:

- **Hydrophone 1:**
  - Capacity: 247pF
  - Sensitivity: 0.35pC/bar
  - Sensitivity in mV/bar: $U_{\text{probe}} = 1.178\text{mV/bar} = 0.01178\text{V/MPa}$

Figure 6.7.: Detail of the silver tip of a needle hydrophone above a multiwell plate, positioned on an optical table. The hydrophone tip oxidizes after a few dozen measurements and regularly has to be replaced to guarantee reliable and reproducible measurements. Photograph is © Martin O. Steinhauser.

- **Hydrophone 2:**
  - Capacity: 241pF
  - Sensitivity: 0.31pC/bar
  - Sensitivity in mV/bar: $U_{\text{probe}} = 1.065\text{mV/bar} = 0.01065\text{V/MPa}$

All of the above mentioned studies that use the experimental principle denoted in Fig. 6.2 have had major deficiencies in the techniques used to characterize the physical conditions in the vessel containing the medium with the cells. Also, in many of these studies, the preparation of reproducible conditions in cell culture was very problematic, introducing many possible sources of errors into the experiment. For example, in some studies the cells were treated with gel and ice (but some were not), or only cell suspensions instead of adherently grown monolayers (as in our experiments) were used, thus reducing the reproducibility of the experiments. The pressure profiles were measured with piezoelectric elements (polyvinylidene fluoride, PVDF) either in form of needle hydrophones or piezoelectric films that are brought in contact to the surface of the cell culture vessel. In both cases, a transfer medium (either water or grease) serves as the acoustic contact to the piezoelectric element. However, shock waves are known to decay rapidly (within micrometers) in liquids and tissue [113, 114, 233]. Thus, the need for a fluid as contact medium may lead to wrong pressure measurements and it would be desirable to determine shock wave properties on a microscopic scale, rather than on a much coarser scale with a PVDF sensor of millimeter dimensions.

A newly developed optical method to determine the pressure profiles is photon Doppler velocimetry [164]. This technique has been used in the Steinhauser lab in a comprehensive study to determine the velocity profile of the bottom of the vessel during shock wave propagation. The measured profiles serve as input for molecular dynamics simulations that allow for computing the pressure fields within the whole vessel with high time resolution [197]. In this way, the local pressure conditions, i.e. the pressure levels directly at the location of the cells during the shock wave experiment can be thoroughly characterized.

6.2. Experiments on laser-induced shock wave destruction of U87 tumor cells

In this section we introduce our very much improved experimental setup for generating laser-induced shock waves in cell cultures compared to the methods used in previous studies, cf. Fig. 6.2. We will proceed to show that our improved setup leads to very reproducible and reliable results with respect to generated pressure profiles. Our method of choice for shock wave generation in tumor cells is based on the ablation of an absorbing material with a pulsed laser [197, 223] as

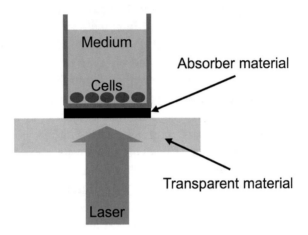

Figure 6.8.: Schematic of our very much improved experimental setup
used to study the effects of laser-induced shock waves on
cells. A pulsed laser beam irradiates an absorbing black varnish at the bottom of the multiwell plate that contains the
adherently grown cells and their growth medium. Shown
here is only one of the wells. At the surface of the absorber,
due to the explosive evaporation of material, a shock wave
is formed that propagates through the bottom into the
vessel. When the varnish is covered with a transparent material (PMMA) facing the laser, the peak pressure of the
generated shock wave is considerably enhanced. Figure adapted from [212].

depicted in the schematic in Fig. 6.8. The bottom of the multiwell plate
is coated with an absorbing black varnish which is applied very thin
in a thorough and elaborate procedure. In addition, the black surface
can be covered with a transparent PMMA (polymethylmetacrylate)
plate which considerable enhances the pressure level of the resulting
shock wave. To generate the laser-induced shock waves, the multiwell
plate is placed on a stage with a hole as shown in the photograph and
the outline in Fig. 6.9. A mirror guides the light of the laser (laser$_G$)
onto the bottom of one well. With this technique, very reproducible

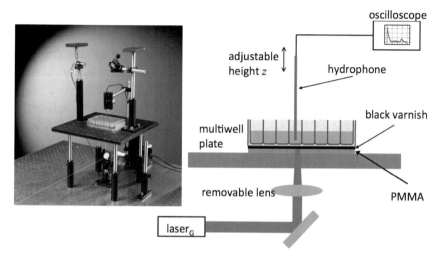

Figure 6.9.: Photograph (©Martin O. Steinhauser) and schematic of our proposed new experimental setup for shock wave generation and pressure characterization in a fluid medium. The pressure curve in the oscilloscope in the figure is an actual pressure measurement.

shock waves are induced that propagate through the absorber material and through the bottom of the well, interacting with the adherently grown cells. A removable lens additionally allows us to focus the beam. To guarantee that the entire well bottom is covered by the laser beam, we use 96-well plates when the focusing lens is in the beam path and 96-well plates when we do not focus the laser. The beam diameter of $laser_G$ is approximately 9mm and the diameter of a well in a 96-well plate is 6 mm. We use a needle hydrophone to record pressure profiles inside the water filled well. The position of the hydrophone z can be adjusted with a motorized actuator.

One general advantage of using a laser for the generation of shock waves over HIFU is the very high reproducibility of the pressure conditions and the spatial precision of wave generation. For the former the pressure profile exhibits no negative pressure contributions, whereas in the latter case the negative part of the pressure profile leads to cavitation effects, which are very hard to control experimentally [48, 49]. In addition, it has been shown, that pressure waves interacting with cells within time intervals of nanoseconds do not lead to any significant increase of temperature [15, 112], which in fact constitutes a major advantage over HIFU. Thus, in contrast to HIFU, any destructive effect, observed as a consequence of the exposure of cells to a laser-generated shock pulses with high gradient, is of purely mechanical nature. This renders laser-induced shock waves an ideal technique to study shock wave effects on cells in cell culture.

Another advantage of our setup is that we use standard culture vessels (48 wells or 96 wells) which are routinely used in cell culture labs. Hence, with our method the cell cultures can be readily analyzed further with standard diagnostic techniques such as confocal or fluorescence microscopy. Furthermore, in our experiments, the pressure threshold for the destruction of human brain tumor cells (U87 glioblastoma) could be determined [197], which – to the best of the author's knowledge – has never been done in any previous studies.

In our study presented here, we systematically analyze our improved experimental setup to investigate the effects of laser induced shock waves on living cells. We use a combined experimental and theoretical approach to characterize the dynamic pressure conditions that the cells are exposed to. In many applications needle hydrophones are used to characterize dynamic pressure conditions in a fluid environment [98, 122]. However, the dimensions of the sensitive element of a hydrophone (diameter of approx. 500μm) are considerably larger than the typical dimensions of biological cells (with diameters of approx. $(15-150)\mu$m.

To overcome this difficulty we use an optical high-speed velocimetry method, introduced by Strand et al. to measure the velocity profile of the surface on which the cells are adherently grown [164]. These profiles serve as basic input for molecular dynamics simulations of the pressure wave propagation through the bottom of the multiwell plate, by which we can determine the pressure conditions on the relevant length and time scales, i.e. on the scale of micrometers and nanoseconds [197].

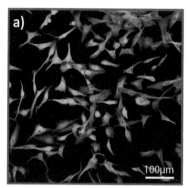

Figure 6.10.: Confocal microscopy images of U87 tumor cells. a) Adherently grown U87 tumor cells. Here, the cells express glial differentiation specific marker glial fibrillary acidic protein (GFAP). b) Trypan blue dying of U87 in a Neubauer chamber for viability testing. Microscopy images are © Martin O. Steinhauser.

6.2.1. Cell culture of U87 glioblastoma cell line

U87, see Fig. 6.10 is a well-established tumor cell line which has been the subject of many biological studies [118]. For example, the U87 genome has been completely sequenced. For the experiments in the Steinhauser lab, U87 was provided by the Department of Neurosurgery at the University Medical Center Freiburg, Germany. U87 were grown in DMEM (Gibco) supplemented with 10% fetal calf serum (FCS) and 1% antibiotics and antimycotics (Gibco).

For the pressure wave experiments, the microtiter plates were prepared as explained above. The cells were cultivated in these plates for 2 to 3 days before the experiments. Tests with non-prepared (transparent) microtiter plates showed that the cells were approx. 80% confluent after this period. When using 96-well plates, 15 wells were exposed to one single laser pulse each. 19 wells on the same plate served as the untreated reference. In experiments with 48-well plates, 6 wells were treated and 16 served as untreated reference. The cell layers were trypsinated with TrypLE (Gibco) and pooled before cell counting without centrifugation. For the cell counting with the hemocytometer, we used a 1:1 mixture of sample liquid with trypan blue. The supernatants were analyzed in the same way.

6.3. Photonic Doppler velocimetry (PDV)

The determination of the velocity profiles of the bottom of the multiwell plates was done by using the technique of photon Doppler velocimetry (PDV), see Fig. 6.11, and was published elsewhere [197]. The reflection of the pressure wave at the well bottom is accompanied by an acceleration of the surface. The pressure of the reflected wave can be derived from the surface velocity profile. PDV is an optical technique that is based on the Doppler shift of the light reflected from a moving object and allows for a non-invasive measurement of velocities with high time resolution. For an enhancement of the PDV signal, the well bottom was covered with reflecting chrome/silver coating or with aluminum foil. Both procedures yield the same results.

The light of a cw (continuous wave) Erbium fiberlaser (referred to as illumination laser) with a maximum output power of is directed onto the surface of the sample by an optical probe. The wavelength λ_R of the laser can be adjusted around the center wavelength of $\lambda_R = 1.557$nm, corresponding to a frequency of f_R $c/\lambda_R = 200$THz (where c is the speed of light). The optical probe basically consists of a lens that focuses the light of the laser onto the surface of the sample. A fraction of the light that is diffusely or specularly reflected by the surface is collected by the same optical probe used for illumination and is separated from the illumination light by a circulator. The reflected light with a Doppler shifted frequency f_D is guided to a high bandwidth optical detector where it is mixed with the light of a second cw laser (referred to as reference laser) with a maximum output power of and with a frequency that is equal to the frequency of the illumination laser f_R. Both lasers exhibit a small bandwidth of less than 1kHz which is essential for a precise velocity measurement.

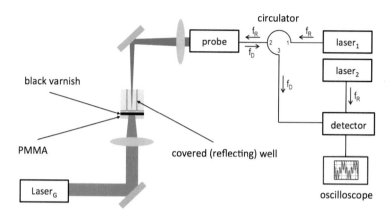

Figure 6.11.: Experimental setup for determination of the velocity profile of the well bottom using PDV. The shock waves are
generated upon illumination of the microtiter plate. For
the velocity measurements we use a high-speed velocimetry system based on the heterodyne method described
in [164]. With this technique the velocity of a moving surface is calculated from the Doppler-shifted frequency f_D
of light that is reflected by the surface. We use two lasers
with a frequency $f_D = 193.414$ GHz. The illumination
laser (laser$_1$) is guided to a probe with an optical fiber and
focused unto the microtiter plate that is covered with a
reflecting chrome/silver coating. The reflected light (with
the Doppler shifted frequency f_D is collected by the same
probe and transported to the detector. At the detector,
the Doppler shifted light and the light of the reference
laser (laser$_2$) are mixed to record the beat frequency f_B of
the resulting signal $I(t)$. To determine the velocity profile
of the well bottom that results from arrival of the shock
wave, we start to record $I(t)$ at the moment of pulse generation with Laser$_G$.

The frequency of the Doppler shifted light that is reflected from the
moving surface is given by

$$f_D = (c + v_s)/(c - v_s)f_R. \qquad (6.2)$$

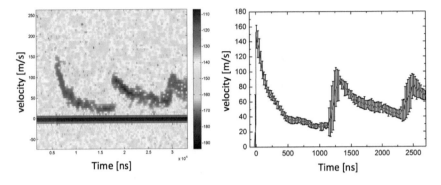

Figure 6.12.: Measured spectrogram of the velocity of the well bottom
(left) and the corresponding $v - t$–diagram (right).

At the detector, heterodyne mixing of the electromagnetic fields of
the Doppler shifted light and the light of the reference laser results in
the formation of a beat signal with a frequency that is equal to the
frequency difference of the superimposed light fields:

$$f_B(t) = 2f_R\big(v_s(t)\big)/c\,. \tag{6.3}$$

The beat signal is recorded with a high bandwidth digital oscilloscope.
For the calculation of the velocity profile, the temporal evolution of
the beat frequency has to be determined. To this end, a Short Time
Fourier Transform (STFT) algorithm similar to that described by
Strand et al. [164] in applied to extract the beat frequency from the
measured signal.

6.4. Results: shock wave damage in U87 tumor cells

To test which peak pressure is needed to kill U87 tumor cells we
use two different techniques to determine the cell concentration: The
Coulter principle and the trypan blue staining method [195]. Since
we are interested in the destructive effect of the shock waves, we carry

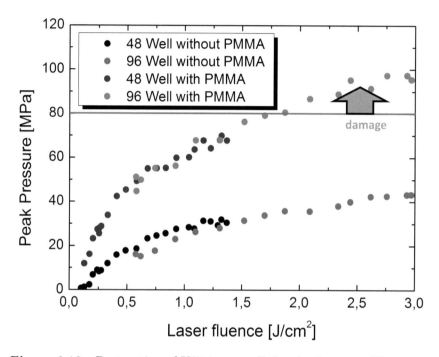

Figure 6.13.: Destruction of U87 tumor cells by shock waves. We extracted the maximum pressure p_{max} of the shock wave for different laser pulse energies at a fixed hydrophone position $z = 2,000 \mu m$. Here we show the peak pressure as function of the laser fluence (pulse energy per unit area). The light and dark dots distinguish data obtained from 96- and 48-well plates. When covered with PMMA, considerably higher peak pressures beyond the critical threshold of 80MPa can be achieved.

out experiments in 96-well plates with PMMA cover. In this setup, we get the highest peak pressures of up to 100MPa and beyond, see Fig. 6.13, that can be tuned with a Q-switch for the laser pulse energy.

Each experiment is carried out in the following way: The cells are grown in the prepared multiwell plates for two to three days. We

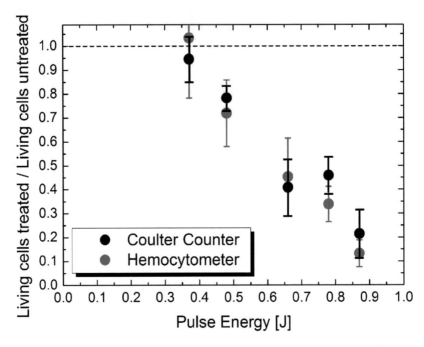

Figure 6.14.: Total number of cells found in the treated sample divided
by the total number of cells found in the untreated sample.
Cells are counted with two different methods as indicated.
All data shown are recorded in 96-well plates with PMMA
cover. Data points are averages of four independent measurements and the error bars are standard deviations.

then expose several wells to single laser pulses. Subsequently, we
determine the number of living cells found in the cell layer (i.e. not in
the supernatant) of a treated well $n_{\text{live}}^{\text{tr}}$ and the number of living cells
found in the cell layer of the untreated well $n_{\text{live}}^{\text{untr}}$. Figure 6.14 shows
the ratio of treated to untreated living cells for different laser pulse
energies. Less than 20% of the tumor cells survive the maximum pulse
energy of $E = 0.87\text{J}$. We also analyze the supernatant in the wells and
determine the number of living and dead cells with the hemocytometer.
It turns out that:

1. most of the dead cells are found in the supernatant, and

2. a large fraction of cells is completely destroyed.

To double check the last point, we use a Coulter cell counter to determine the volume distributions of the particles in our multiwell plates. In Fig. 6.15 a)-c) we display a schematic of the principle of a Coulter counter along with the results of cell counting in the treated sample (blue and red lines) and in the untreated reference (green and black lines). Each curve is an average of five measurements at the same sample. Note the comparatively small error bars which are displayed as filled areas, denoting the standard deviations. The pressure wave was generated at the full laser pulse energy of of our laser system with $E = 0.87J$ per pulse in a 96-well plate with PMMA cover.

We analyzed the volume distributions measured by a Coulter counter of a sample treated at the maximum laser pulse energy, see 6.15. For comparison, the green and the black lines show the volume distributions of the cell layer and the supernatant of an untreated reference sample, respectively. We find that there are hardly any particles found in the untreated supernatant (black curve). The volume distribution of the untreated cell layer (green curve) also has the expected form with a mean volume of 1.75pl. This volume corresponds to an approximate cell diameter of 15μm, which is in agreement with the microscopic images of U87 cells shown in Fig. 6.10. In the treated sample, the volume distribution of the cell layer (blue line in Fig. 6.15) has the same form as the untreated cell layer but with a smaller total number of cells. The supernatant of the treated sample (red line in Fig. 6.15) on the other hand shows a large peak of small volume which is cell debris from completely destroyed cells. Hence, the volume distributions obtained from the cell counter confirm that the cells are destroyed and detached from the bottom of the vessel by the shock wave.

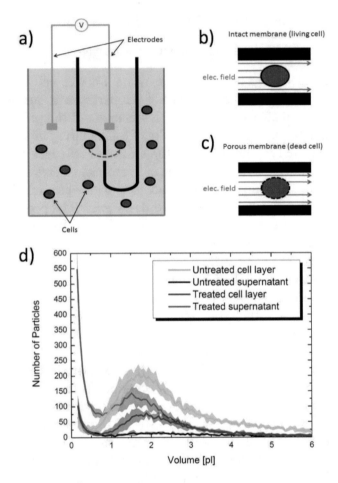

Figure 6.15.: a)-c) Coulter principle of cell counting: The instrument probe, which is placed in a vial of cells, contains a small hole or aperture through which the cells can pass. There are electrodes on either side of the aperture, such that a current flows across them. As cells are drawn through the aperture, they momentarily block the current, producing an electrical spike that the instrument detects. The amplitude of the pulse is proportional to the 3D volume of the cell. d) Volume distributions of our adherently grown cell layer and of the supernatant with standard deviation.

6.5. Simulation of shock wave damage in coarse-grained models of membranes

In this section we present shock wave simulations of CG models of lipid bilayer membranes using our proposed CG model of Chap. 5 and our new multiscale coupling scheme presented in Chap. 3. Figure 6.16 shows a simulation snapshot of a model membrane (made up of CG lipid molecules) along with its surrounding water (represented as a continuum, i.e. as discrete integration points, misleadingly called SPH-particles).

CG models are nowadays routinely used in polymer- and biophysics, mostly for equilibrium [26, 63, 69, 88, 173, 207, 250], but also for non-equilibrium studies [82, 109, 181] of lipid bilayer membranes and provide a description of reduced complexity with respect to the molecular degrees of freedom [170, 184, 238]. Almost all of the existing body of simulation studies of biomembrane properties have been performed at or very near at equilibrium.

However, the exploration of the interaction of shock waves with biomembrane structures constitutes a non-equilibrium physical process. In the few existing computational studies exploring shock waves in soft matter none takes properly into account the particular physical conditions of a shock wave and they are limited to unrealistically small systems with explicitly modeled solvent [125, 126, 138]. Even in the largest existing all-atom simulation published to date [125, 126], the size of the considered system was several orders of magnitude too small in size and too limited with regards to the time scale to be able to capture any relevant effects of shock waves in membranes observed in experiments, such as transient permeability and subsequent self-repair of parts of the membrane in an eukariotic cell [82].

It is well known that the exposure of biological cells to shock waves can cause damage of varying extend to the cell membrane [59, 61, 122, 141] depending on the pressure level of the shock wave [197, 223]. However, the exact mechanisms by which shock waves interact with membranes of biological cells is vastly unknown which is in

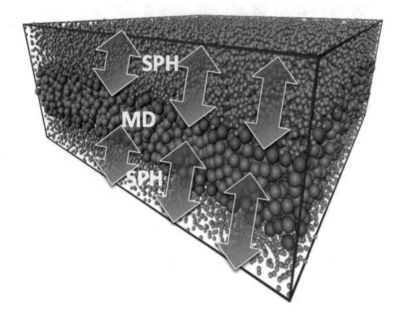

Figure 6.16.: Snapshot of an equilibration state of our CG membrane model using MD/DPDE particles and water as the surrounding SPH continuum [221]. The computational coupling of the CG membrane and the surrounding continuum is indicated with arrows. The simulations were done with the author's software suite `MD-CUBE`.

striking contrast to the widespread use of shock waves e.g. in medical treatments. Here, the use of shock waves as a form of therapy has two major areas of application:

1. Gene therapy: Here, the plasma membrane is transiently punctured by the treatment with shock waves, which can be used for delivering genetic material or therapeutic molecules into the cytoplasm [121].

2. High Intensity Focused Ultrasound (HIFU): Here, tumor tissue is treated with HIFU which transports energy to a focal point where

it is converted into heat, leading to denaturation of proteins in
the cells (coagulative necrosis). A secondary and hard to control
destructive effect during HIFU treatment is caused by acoustic
cavitation, i.e. the formation and implosion of small gas bubbles at
the focal point.

Experimentally, direct visualization of the dynamics of membrane
rupture is very difficult to achieve, because a typical cell membrane
has a thickness $d \approx 5$nm, while the pressure front of a shock wave
travels at supersonic speeds, i.e. with a velocity $v_s \gg 1,430$ms^{-1} in
water. The time-scale during which a shock wave interacts with a cell
membrane is thus on the order of a few picoseconds.

The structural complexity of cell membranes adds on to the computational challenges involved with simulating CG models of membranes:
They are constituted mainly from phospholipids, which form a two-
dimensional bilayer, with the hydrophilic phosphate „heads" pointing
outwards and interfacing with the aqueous environment, while the
hydrophobic „tails" point inwards in order to minimize their exposure
to the surrounding water. This structure is essentially a thin fluid layer
with immersed proteins and carbohydrates, cf. Fig. 5.5, stabilized by a
subtle balance of competing hydrophobic and hydrophilic interactions.

Studying the interaction of a shock wave with a *soft* matter system
such as a lipid bilayer is interesting in its own right: as the lipid bilayer
is not a crystalline solid, the shock wave does not experience a major
change in impedance as it traverses the interface between water and
bilayer, and it is *a priori* unclear by which mechanisms damage is
caused, and how it depends on physical parameters such as shock wave
velocity, shock pulse duration, or shock pulse shape.

The standard technique to numerically simulate phospholipid bilayers is MD [10, 155, 189, 210]. However, in spite of considerable
improvement of the method itself and the ever-increasing CPU power
over the last decades, it is still only possible to simulate – with atomistic detail – tiny membrane patches, typically involving only a few
hundred lipid molecules for time spans of a few hundred nanoseconds.
The existing published research on the *largest atomistic* MD simula-

tions of shock wave interaction with cell membranes that was done by Koshiama et al. [125, 126] uses system sizes of 128 phospholipids, with simulation box lengths parallel, and perpendicular to the bilayer plane of $L_\parallel \approx 6.5$nm and $L_\perp \approx 16.0$nm, i.e., comparable to the bilayer thickness. However, it is very questionable whether such small systems can quantitatively reproduce real damage processes because:

- The employed periodic boundary conditions impose an artificial stabilization of the membrane patch due to correlation effects.

- Membrane rupture will likely originate from a defect, i.e. a deviation from the ideal flat surface. However, undulations of the membrane are strongly suppressed due to the small simulation box.

- The time evolution of the shocked membrane needs to be studied for as long as possible, requiring a large box length $L_\perp$ along the direction of the shock impulse in order to allow the shock front to travel for a long period of time.

The last point is important because either a slow, diffusive disintegration of the membrane, or structural recovery from the induced damage can be observed. Assuming the validity of Moore's law in the future (which yet is no longer valid today), we could in principle wait until the computers are powerful enough to simulate shock wave damage in a membrane which has the size of a typical human eukariotic cell, i.e. $\approx (30 - 50)\mu$m and then simply upscale Koshiyama's or some other atomistic model similar to that. However, doing the algebra leaves us with a time span in the best of all cases of at least 40 years. Thus, mere brute force computational speed is not a useful option – at least not for the current and next generation of researchers. This is why we have to come up with more clever ways to bridge the length scales from the nano to the micro regime and beyond.

6.5.1. Propagation of the shock wave

Using the DPDE thermostat [70], we perform shock wave simulations of square lipid bilayer patches for values of the lipid interaction parameter in the interval $h_c \in [0.8, 1.4]\,\sigma$. For $h_c \lesssim 0.7\sigma$, the bilayer is instable and dissolves, while it is crystalline for $h_c \gtrsim 1.5\sigma$. Thus, the investigated range is representative of a wide spectrum of lipid bilayers with different mechanical stability. Initial configurations are taken from equilibrium runs performed in the tensionless state. The considered system here consists of $N_{\mathrm{lip}} = 4.000$ lipid molecules, with $L_\| \simeq 20.0$ nm. The simulation box length along the direction normal to the bilayer is chosen as $L_\perp = 80.0$nm. The total number of particles (CG particles treated with MD plus SPH particles of the continuum) is $N \approx 1.2 \times 10^6$. The velocity of the piston that generates the shock wave is varied in the range $v_p \in [2,000, 6,700]\,\mathrm{ms}^{-1}$ in order to produce shock waves with different supersonic velocities. After 164 simulation time steps of $\Delta t = 20.0$fs, the piston is stopped, while the initiated shock wave continues to propagate further along $L_\perp$. The lipid bilayer is placed 9.0nm in front of the initial piston position, far enough away not to be hit by the piston. Snapshots of an exemplary simulation are shown in Fig. 6.17, where the shock wave was initiated by a piston with velocity $v_p = 450$ms^{-1}. In the figure, it can be seen that the lipid bilayer has a momentum in positive x-direction (to the right) during the simulation due to the momentum-transfer by the piston.

In Fig. 6.18 we show two simulation snapshots displaying the SPH particles which are modeled using an equation of state for H_2O, surrounding the lipid bilayer membrane. Membrane molecules are color-coded in red (hydrophobic head part), green (hydrophilic tail) and one layer of water molecules is displayed in blue. Note that after the shock wave has passed the membrane water molecules penetrate the lipid bilayer the structure of which is destroyed.

Precise visual inspection of the lipid bilayers after shock wave impact in Fig. 6.19 shows that the here considered range of shock front speeds causes little to heavy damage (complete destruction) to the equilibrium bilayer structure.

For a thorough analysis of shock wave experiments, it is necessary to follow the shock front velocity and peak pressure, as it travels through the simulation box and dissipates. Consequently, we discretize the

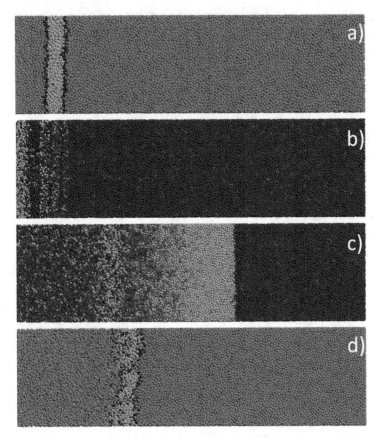

Figure 6.17.: Snapshots of a shock wave simulation with a lipid bilayer
in a shock tube at different times.The simulation box
measures $(80.0 \times 20.0 \times 20.0)\,\mathrm{nm}^3$, comprising 1.2×10^6
particles. A piston, moving from left to right with velocity
velocity of $4,500\mathrm{ms}^{-1}$ compresses the material ahead of
it and induces a shock wave. a,b) and c,d) correspond to
simulation times of 0.8ps and 14.0ps, respectively. In a)
and d) the particles are colored according to type (blue:
H_2O, red – lipid heads, green – lipid tails). In b,c) the
pressure distribution is shown, with blue and red signifying
low and high pressure, respectively. Peak pressures in b)
are 31.2GPa and 0.9GPa in c), respectively.

a)   b)

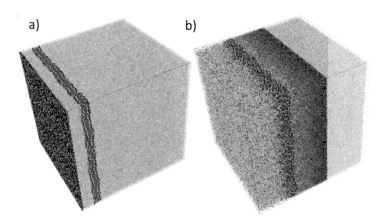

Figure 6.18.: Shock wave simulation of our membrane-fluid system displaying the amphiphilic lipid molecules forming the bilayer membrane and the fluid (SPH) particles [221]. Simulation with 1.2×10^6 particles. a) The system just after the shock wave has been initiated, propagating from left to right. b) The system when the shock wave has passed the membrane. One can see a rather complete destruction of the membrane.

simulation volume in thin slabs of width 0.5nm which are oriented parallel to the yz-plane. By computing the average velocity and pressure in each slab for each simulation timestep individually, the evolution of the shock wave front can be studied in detail. The pressure profiles of the systems as displayed in Fig. 6.17 are shown in Fig. 6.21: at $t = 0$ps, one can identify the location of the lipid bilayer by the interfacial pressure variations near $x \simeq 9.0$nm. The profiles at $t = 1.0$ps and $t = 14.0$ps clearly show the shock wave front with an associated pressure peak. The fast dissipation of the shock wave is demonstrated by the quickly diminishing peak pressure heights. The minor pressure peaks near $x \simeq 19$nm and $x \simeq 22$nm in the $t = 12.5$ps and $t = x24.0$ps profiles, respectively, are due to the phospholipid bilayer, which has moved to the right during the simulation, as the compressing piston transfers a linear momentum in

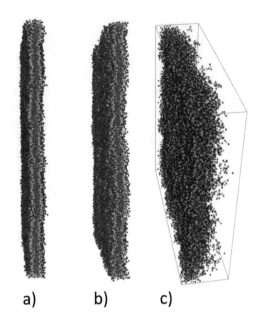

a)   b)   c)

Figure 6.19.: Visualization of the damage caused in the bilayer due to
shock waves with different shock front speed v_s. Snapshots are shown for the system of Fig. 6.18. SPH particles
of the continuum have been left out for better visualization of the bilayer. a) initial membrane at equilibrium;
b) $v_s = 500\text{ms}^{-1}$, where the membrane in essence stays
intact after shock wave impact; c) $v_s = 5,100\text{ms}^{-1}$, with
total destruction of the membrane's lipid bilayer structure.

x-direction. Interestingly, the bilayer retains a relatively high pressure
after the shock wave front has passed over it. This can be explained
by noting that, in contrast to the water particles, the CG phospholipid
molecules are modeled with harmonic springs, which can store energy
in oscillatory motions.

For determining the speed of the shock wave, we identify the x-
location of the shock wave front, $x_s(t)$, as a function of simulation
time. A simple linear function, which describes the attenuation from
the hypersonic velocity to constant sound speed is then fit to these

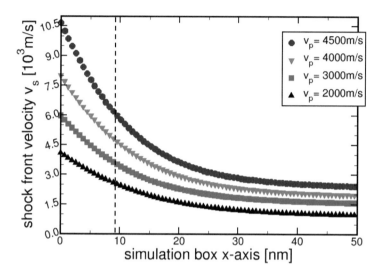

Figure 6.20.: Shock front velocity v_s as a function of different piston velocities v_p. The dashed line indicates the location of the lipid bilayer at $x = 9.0$nm.

location data:

$$x_s(t) = \frac{\alpha t}{t + \beta} + ct. \qquad (6.4)$$

This saturation-like function fits the recorded data very well, with the fitted value of c being close to the sound speed of the medium, as is physically required. The speed of the shock wave front is given by the derivative of $x_s(t)$ with respect to time,

$$v_s(t) = \frac{\mathrm{d}x_s(t)}{\mathrm{d}t} = \frac{\alpha}{\beta + t} - \frac{\alpha t}{(\beta + t)^2} + c. \qquad (6.5)$$

As the precise location of the bilayer is known, the time of impact is given by solving (6.4) for t, and calculating the impact velocity from (6.5). This approach has the advantage that it yields an analytic expression for the velocities. Calculating the velocity directly from finite time differences of the location data is not a viable alternative due to numerical noise. Figure 6.20 shows the thus obtained velocities

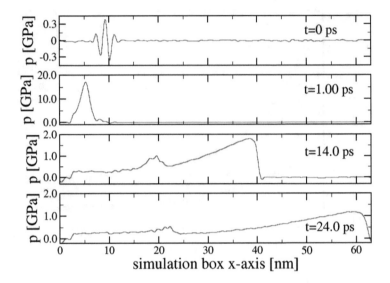

Figure 6.21.: Pressure profiles at different times resulting from an initial piston velocity of $v_p = 4,500\text{ms}^{-1}$. The system is the same as described in Fig. 6.17. Note the different ordinate scales.

as a function of distance from the compressing piston. It is clear from this graph, that attenuation of the shock wave front velocity proceeds very quickly, in agreement with experiment [201].

6.5.2. Membrane damage: membrane order parameter

There are a number of different phenomena associated with lipid bilayer damage:

- The bilayer can tear along well localized paths leaving the majority of its area intact.

- The two bilayer leaflets can inter-penetrate upon compression.

- Single lipids can move out of their equilibrium positions and orientations via diffusive mechanisms.

We use a pragmatic approach to combine all of these effects into a single scalar order parameter, Ψ, which is based on the projection of the particle pair distribution function on rotational invariants [95]. To do so, we define the orientation dependent correlation function

$$C(r) = \frac{1}{4\pi N} \sum_{i>j}^{N} \delta(r - r_{ij}) \frac{\vec{e}_i \cdot \vec{e}_j}{r^2}, \qquad (6.6)$$

where the sum runs over all pairs of lipids i and j, r_{ij} is the pair distance as measured between the central beads, and $\vec{e}_i$ and $\vec{e}_j$ are orientation vectors, i.e. the normalized distance vectors between the first and the third bead of each lipid. $C(r)$ is similar to the familiar radial distribution function, except that it is additionally weighted by relative orientations. To convert this distance dependent correlation function into a single scalar, we integrate $C(r)$ up to a distance $r_c = 1.0$nm, which is chosen such as to include the first coordination shell, i.e., the first peak in $C(r)$, in the undamaged bilayer structure:

$$\Psi = 4\pi \int_0^{r_c} \mathrm{d}r\, C(r) r^2 \,. \qquad (6.7)$$

Disruption of the equilibrium bilayer configuration affects this order parameter in two different ways: If the mutual orientation of neighboring lipids deviates from parallel alignment, or if lipids are separated from each other beyond their equilibrium distance r_c, the value of Ψ is reduced, thus providing a quantitative means to characterize lipid bilayer damage.

6.5.3. Membrane damage: effects of shock wave speed and system size

Our CG model of polymers introduced in Chap. 5 is capable of simulating lipid bilayers of different stability as displayed in the calculated phase diagram of Fig. 5.7. Possible phases of the bilayer range from an unordered, instable gas phase, a transition state to stable bilayers

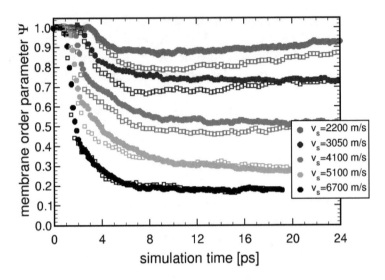

Figure 6.22.: Effect of lipid attraction parameter h_c on bilayer damage induced by shock waves of different incident velocities v_s. Filled circles exhibit results for the CG lipid model with interaction parameter $h_c = 1.4\sigma$, being representative of a very stable lipid bilayer. Open squares show the induced damage for lipids with $h_c = 0.8\sigma$, which is close to the lower end of the stable bilayer regime in the phase diagram in Fig. 5.7.

and vesicles to almost solid-like bilayer structures (gel-sol transition) with strong peaks that can be seen in the pair correlation function of Fig. 5.10. Here, we exploit this feature of our CG polymer membrane model, in order to study the relation of bilayer stability to shock wave induced damage.

Our choice of system size here is $N_{lip} = 10^3$ lipids, with a lateral simulation box length $L = 80$nm, which is large enough to capture the characteristic bilayer undulations driven by thermal fluctuations. We verified this in Fig. 5.11, where we show that the Fourier transform of the lipid bilayer height fluctuations already reach the asymptotic scaling regime predicted by continuum theory. Hence, our choice of

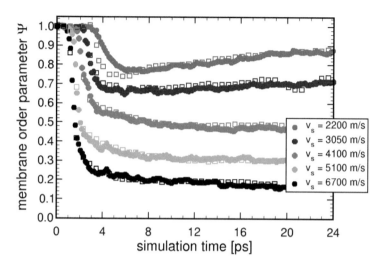

Figure 6.23.: Membrane damage induced by shock waves of different incident velocities v_s. Simulation data are presented from top to bottom with increasing shock wave speed for the CG lipid model with interaction parameter $h_c = 1.0\sigma$ and two different system sizes: filled circles and open squares are for bilayers with 10^3 and 10^4 lipids, respectively.

lipid bilayer area is of a size, representative of a real membrane and the periodic boundary conditions applied in the numerical simulation do not introduce artificial errors.

Figure 6.22 compares Ψ of two bilayers with $h_c = 0.8\sigma$ and $h_c = 1.4\sigma$, which are representative of the extreme ends of the stable bilayer range, namely close to the gas phase and close to the crystalline phase, cf. Fig. 5.7. The graph shows a clear difference in the resulting damage of these two systems, however, it is surprising that bilayer stability leads to a quantitatively small effect, especially at long times after shock wave impact. A possible explanation for this result is that the difference in cohesive energy density between these two systems is smaller by orders of magnitude than the energy densities introduced temporarily by the shock wave.

Finally, in Fig. 6.23 we show the damage caused by shock waves of different impact velocities v_s for membrane systems of medium stability with an interaction parameter $h_c = 1.0\sigma$. As one can see, higher impact velocities cause a stronger decrease of Ψ, i.e. larger damage. The initial descent of Ψ is almost instantaneous for the here simulated shock front speeds, however, for large simulation times Ψ increases within a few ps, indicating reversible damage. This is consistent with the experimentally observed self-repair of biomembranes that are transiently permeable when exposed to shock waves [159]. For small shock front velocities $v_s \lesssim 3,050\text{ms}^{-1}$, no such clear increase of Ψ is observed. Thus, the existence of a critical shock front velocity for self-recovery of membranes can be deduced from our calculations.

7. Final considerations

In this work, important novel computational aspects of multiscale modeling methods and their application to shock wave phenomena in granular solids, ceramics, polymers and bilayer membranes are presented. Additionally, the results of corresponding high-speed impact experiments with ceramics are shown, along with laser-induced shock wave experiments with U87 glioblastoma (human brain tumor) cells. We successfully tested the hypothesis that mechanical shock waves can have a destructive biological effect on this type of tumor cells. Here, we also show a state-of-the-art realization of photon Doppler velocimetry which was used to measure the velocity of the bottom surface of the cell culture vessels exposed to shock waves. In summary, in this work we discuss or introduce:

- a multiscale method for coupling the atomic (MD/DPDE) with the continuum domain (SPH),

- a discussion of multiscale modeling with major methods on the various length scales,

- a new idea of how to generate realistic microstructures of granular materials in 3D by means of power diagrams,

- a particle-based multiscale model which allows for simulating a macroscopic specimen size while including microstructural features,

- a CG model for both, polymers and phospholipid molecules,

- crossover scaling of semiflexible polymers,

- a phase diagram and other equilibrium properties of a CG polymer model for lipid bilayer membranes,

© Springer Fachmedien Wiesbaden GmbH 2018
M. O. Steinhauser, *Multiscale Modeling and Simulation of Shock Wave-Induced Failure in Materials Science*,
https://doi.org/10.1007/978-3-658-21134-9_7

- new and high-precision experiments on laser-induced shock wave destruction of human U87 glioblastoma brain tumor cells,

- simulations and detailed investigations of shock-wave induced damage in biomembranes.

We also demonstrate with this work the advantages of a *combined* experimental-computational research approach utilized in the Steinhauser lab, which is essential for the achievement of a deeper understanding of material behavior on multiscales. For example, all of the presented shock wave experiments on ceramic materials and on U87 tumor cells have been directly performed by or under the supervision of the author of this work.

7.1. Shock wave physics and multiscale modeling

In the first part of this work, theoretical-numerical foundations for shock wave simulations are covered and the shock tube as a numerical or experimental tool for shock wave research is introduced. We discuss the importance and abundance of multiscale methods in materials science. A classification of different applications of multiscale methods and an overview of the major simulation methods on different length scales is provided. Along with multiscale methods, computer simulation has fully emerged as a research tool that complements experiment and theory. After introducing the research codes we use to generate the results discussed in this work, we present a new thermodynamic method for coupling the atomic scale with a continuum description of condensed matter. The method uses the DPDE thermostat to transform kinetic particle energy from the atomic domain into the internal energy reservoir of the continuum domain. We implement our coupling scheme with the MD method for the atomic scale simulation and the SPH method for the continuum. We show the working of our new coupling scheme by applying it to a shock tube problem where we have separate regions of MD and SPH particles. To the best of our knowledge, with these simulations we are the first ones to apply the DPDE method to a useful problem of physical chemistry. The great promise of this new coupling scheme is a tremendous gain in simulation time since the fluid surrounding a soft matter system, e.g. a biomembrane, can now be simulated as continuum, thus reproducing the correct hydrodynamic behavior with less particles than were necessary to simulate the fluid *explicitly.*

7.2. Multiscale modeling of granular matter

In Part II of this work, a new method for generating three-dimensional microstructures of granular materials is presented which is based on a mathematical extension of the Voronoi diagram by introducing additional weighting points in the diagram, thus turning it into a PD. The theory allows for calculating the diagram in 3 dimensions with worst-case optimal algorithms that scale with $N \log N$ in 3D. In essence, the PD is a 3D polyhedral section of a PD in 4D and can be calculated efficiently. Polyhedral structures are the predominant form of the microstructures of ceramics, so we believe that this method is a very appropriate way to generate microstructural features in 3D. With a Monte-Carlo scheme we optimize these structures with the only available 2D data from experimental micrographs. We present FE simulations that are based on the microstructures generated with our new method and discuss the principal limitations of FE simulations when trying to incorporate microstructural details in a purely continuum-based computer code.

We also introduce a new particle-based multiscale method for the numerical investigation of crack initiation and propagation, impact-induced shock waves and failure in granular materials. Despite the simplicity of the model with only three free parameters to be fitted to mechanical values of a material specimen we see good agreement with high-speed impact experiments which we perform in EMI labs and which serve as test cases for the numerical simulations performed with the author's software suite MD-CUBE.

The application of shock waves in hard matter systems reveals that the fracture processes can be understood already with relatively simple parametric models based on networks of non-entropic springs which capture the essential features of a material with respect to pressure and tensile loading conditions. One advantage of our proposed particle model in contrast to FE models is, that *many* systems with the same set of parameters (and hence the same macroscopic properties) but statistically varying microstructure (i.e. initial arrangement of particles) can be simulated, which is very awkward, if not impossible to attain by using the FEM. By way of performing many simulations, a good statistics for corresponding observables can be achieved.

7.3. Multiscale modeling and shock wave simulations of soft matter

In Part III of this work, a CG model of macromolecules is presented and tested against the predictions of scaling theory for the crossover scaling of linear polymers when undergoing a transition from fully flexible to semiflexible. It is shown that with CG models the scaling laws of polymer theory can be recovered. The simulation studies reported so far in the literature were based on detailed atomistic models and have failed to show the expected p^{-4} scaling of $\langle X_p^2 \rangle$ and of τ_p. By way of comparison with the data presented in our MD simulation study, simulating single polymers and polymer melts with larger persistence lengths L_p than in the previous numerical studies, we were able to reveal the reason for this failure: The persistence lengths L_p in all previous studies were too small in order to enter the p^{-4} scaling regime of bending modes. All previous numerical works were actually limited to the crossover scaling regime, which delivers an easy explanation why in those studies it was wrongfully speculated on an apparent $p^{-2.3}$ and $p^{-2.6}$ scaling [34], respectively a p^{-3} scaling [132, 180] in the bending mode regime.

The data in our MD study reveal the existence of only two clearly distinct scaling regimes where τ_p scales according to p^{-2} (Rouse scaling), respectively p^{-4} in the regime that is dominated by bending modes. It is shown that these two regimes can be described very well with our proposed analytical model which accounts for both scaling regimes as a result of an entropic force term and a bending force term that are both needed to describe semiflexible behavior.

In the last chapter of Part III, experimental research on the shock wave destruction of human brain tumor cells, U87 glioblastoma, is presented. Here, we show considerable improvement of the experimental setup to render the biological experiments very reproducible with small statistical errors. We determine for the first time the shock pressure levels necessary to achieve a biological destructive effect and show convincingly that beyond that threshold, laser-induced shock waves are able to destroy U87 tumor cells. Finally, the introduced CG and multiscale methods are applied to the shock wave simulation of phospholipid membranes using our new multiscale coupling scheme and the previously introduced CG model of lipid bilayers.

7.4. The simulation of shock waves in membrane structures

In this work, we present computer experiments on the interaction of shock waves with coarse grained models of lipid bilayers, which are the dominant constituents of the plasma membrane and the endomembrane in eukaryotic cells. The study of shock wave phenomena in soft matter systems is still in the early stages as a field of research and in fact, the combination of a CG description of polymers with shock wave phenomena has been a quite new approach that emerged only recently in key publications by Steinhauser and co-workers, see e.g. [82, 212, 213, 223]. Hence, there are currently very few other publications to be found literature, in which this combined coarse-grained/shock wave approach is employed.

Our model, which represents an entire lipid molecule with a number of particles, aims at computational efficiency rather than capturing molecular detail. This way, we can simulate larger systems within reasonable CPU time than what is currently possible with atomistic MD simulations. Our simplified model successfully reproduces key mechanical properties of real lipid bilayers such as the bending rigidity, the fluid character, and the area compression modulus. The lack of molecular detail is a particular advantage for the presented study, as it aims to elucidate the basic damage mechanisms when a soft structure such as a lipid bilayer is impacted by the energy of a shock wave.

The lipid bilayers in our computational studies have a lateral size of up to $L \approx 70\text{nm}$. This system size is large enough to accommodate the characteristic bilayer bending modes, which can be considered a minimum requirement in order to rule out finite-size effects. This system size, together with the large amount of water required to model the aqueous environment would have required enormous computational resources if full atomistic MD had been used instead of our CG description. The here employed DPDE technique facilitates faithful simulations of non-equilibrium processes such as shock waves, as it allows to recover the correct heat capacity for coarse grained models with highly reduced number of degrees of freedom.

The parameter range of the lipid interaction potentials is chosen such that mechanically barely stable bilayers as well as very stable bilayers can be simulated. Therefore, our numerical experiments are representative of a wide range of different real phospholipid bilayers. The results of our shock-wave experiments indicate a threshold shock front velocity roughly at $v_s \approx 3.050\text{ms}^{-1}$, below which the bilayer recovers from shock wave induced damage on the timescale of a few ps as reported in literature [159]. Above this velocity, no such recovery could be observed and the damage was irreversible. Most of the damage is caused instantaneously while the shock-wave front traverses the lipid bilayer, with only little more damage occurring afterwards. A strong correlation between v_s and the induced damage is observed, however, the dependence of the induced damage on the lipid attraction parameter h_c is found to be weak. Different mass distributions within the lipid bilayer are found to have no effect, thus ruling out inertial effects for the damage mechanism.

In comparison with two publications on the interactions of shock waves with lipid bilayers by Koshiyama et al. [126, 125], which are both detailed all-atom simulations, we point our the following: When using a CG model, one is able to study much larger systems with a membrane area that is multiple times larger than in the largest atomistic simulations. Furthermore, one can to increase the simulation box length parallel to the shock front direction. This is an important detail, as it enables one to simulate the bilayer in a state which is not affected significantly by finite-size effects. Additionally, due to the increase in box length, one can study the damage evolution for much longer simulation times: the afore mentioned atomistic studies were limited to 600 fs, while we extended this time well into the picosecond range. In terms of quantifying the induced damage, Koshiyama et al. mainly employed the bilayer thickness, which temporarily decreased during compression, and the Deuterium order parameter [67], which is an accumulated orientational order parameter similar to the one we introduced in (6.7). Their results for a single DPPC system are qualitatively similar to the results presented here, however, in contrast to their work we have studied a much wider range of bilayers of different stability. In addition, we have identified a threshold shock-front velocity below which the induced damage is reversible on the timescale of a few ps. In summary, when comparing our CG model of lipid bilayer membranes with detailed all-atom simulations, we conclude that we have extended the time and length scale that can be treated in computer simulations of such soft matter systems.

7.5. What is the future of multiscale materials modeling?

The importance of materials modeling across length and time scales will further increase in the future with the development of new computational techniques and the advent of exascale computers. Currently, mostly sequential multiscale modeling and bridging of few scales is state of the art, while concurrent modeling concepts begin to emerge.

It can be expected that the future will bring more concurrent modeling for a variety of materials that would enable design of materials at the macroscale by using an inverse engineering design methodology. Concurrent modeling can be achieved through further development of a systematic coarse-graining methodology that retains microstructure heterogeneities and chemistry of the system at higher scales while the reverse process will utilize general backmapping algorithms that can uncover underlying atomistic structure and local heterogeneity from the high-level CG representation.

New algorithms will likely improve the efficiency and validity of computations at every scale by making the representation of the total material system more accurate with better estimates of the uncertainties that have arisen. For example, embedded and adaptive methods may enable more accurate calculations of the material system at the various scales (quantum, atomistic, micro- and mesoscales) subjected to a dynamically adjusted environment. Current, predetermined a priori representation of the multiscale material system for atomistic simulations (force fields) or for FE techniques (constitutive equations) could be calculated on the fly, thus adding accuracy and less uncertainty to the material system simulation.

Undoubtedly, with increased coupling of different algorithms across spatiotemporal scales, understanding deficiencies and the range of applicability of models, as well as estimation of error on simulations, will become of vital importance. The field of verification, validation, and uncertainty quantification will become an integral part of concurrent multiscale modeling. In the future, we can expect an increasing integration of multiphysics within multiscale modeling, allowing for realistic simulations under various external fields and extreme conditions. Bulk calculations might be replaced with realistic simulations including local heterogeneities, such as stochastically distributed defects, grain and interface boundaries or microphase-separated morphologies that can dynamically evolve upon influence of external fields.

Bibliography

[1] F. F. Abraham, D. Brodbeck, R. Rafey, and W. Rudge, Instability dynamics of fracture: A computer simulation investigation, *Phys. Rev. Lett.*, 73 (1994), 272–275.

[2] F. F. Abraham, D. Brodbeck, W. E. Rudge, J. Q. Broughton, D. Schneider, B. Land, D. Lifka, J. Gerner, M. Rosenkrantz, J. Skovira, and H. Gao, Ab initio dynamics of rapid fracture, *Model. Simul. Mater.*, 6 (1998), 639–670.

[3] F. F. Abraham, J. Q. Broughton, J. Q. Broughton, N. Bernstein, E. Kaxiras, and E. Kaxiras, Spanning the length scales in dynamic simulation, *Comp. Phys.*, 12 (1998), 538–546.

[4] F. F. Abraham and H. J. Gao, How fast can cracks propagate?, *Phys. Rev. Lett.*, 84 (2000), 3113–3116.

[5] M. Abramovitz and I. A. Segun, *Handbook of Mathematical Functions*, Dover Publications, New York (1964).

[6] B. Alberts, D. Bray, A. Johnson, J. Lewis, D. Morgan, M. Raff, K. Roberts, and P. Walter, *Molecular Biology of the Cell*, Garland Science, Taylor and Francis Group, New York, 4 edn. (2000).

[7] B. J. Alder and T. Wainwright, Molecular dynamics by electronic computers, in I. Prigogine (editor), *Proceedings of the International Symposium on Transport Processes in Statistical Mechanics, Brussels, 1956*, pages 97–131, Interscience Publishers, Inc., New York (1958).

[8] B. J. Alder and T. E. Wainwright, Phase Transition for a Hard Sphere System, *J. Chem. Phys.*, 27 (1957), 1208–1209.

© Springer Fachmedien Wiesbaden GmbH 2018
M. O. Steinhauser, *Multiscale Modeling and Simulation of Shock Wave-Induced Failure in Materials Science*,
https://doi.org/10.1007/978-3-658-21134-9

[9] B. J. Alder and T. E. Wainwright, Phase transition in elastic disks, *Phys. Rev.*, 127 (1962), 359–361.

[10] M. P. Allen and D. J. Tildesley, *Computer Simulation of Liquids*, Clarendon Press, Oxford, UK (1987).

[11] J. R. Asay and L. C. Chhabildas, Paradigms and Challenges in Shock Wave Research, in *High-Pressure Shock Compression of Solids VI*, pages 57–119, Springer New York, New York, NY (2003).

[12] F. Aurenhammer, Power Diagrams: properties, algorithms and applications, *SIAM J. Comput.*, (1987).

[13] J. B. Avalos and A. D. Mackie, Dissipative particle dynamics with energy conservation, *Europhys. Lett.*, 40 (1997), 141–146.

[14] G. S. Ayton, W. G. Noid, and G. A. Voth, Multiscale modeling of biomolecular systems: in serial and in parallel, *Curr. Opin. Struct. Biol.*, 17 (2007), 192–198.

[15] K. G. Baker, V. J. Robertson, V. J. Robertson, and F. A. Duck, A review of therapeutic ultrasound: biophysical effects., *Phys. Ther.*, 81 (2001), 1351–1358.

[16] P. Ballone, W. Andreoni, R. Car, and M. Parrinello, Equilibrium structures and finite temperature properties of silicon microclusters from ab initio molecular-dynamics calculations, *Phys. Rev. Lett.*, 60 (1988), 271–274.

[17] K. J. Bathe, *Finite Element Procedures in Engineering Analysis*, Prentice Hall, Cambridge (1982).

[18] G. Ben-Dor, T. Elperin, and O. Igra (editors), *Handbook of Shock Waves*, vol. 2 of *Volume 2: Shock Wave Interactions and Propagation*, Academic Press, San Diego, San Francisco, New York (2001).

[19] G. Ben-Dor, Igra, and T. Elperin (editors), *Handbook of Shock Waves*, vol. 3 of *Volume 3: Shock Waves in Chemical Reactions and Detonations*, Academic Press, San Diego, San Francisco, New York (2001).

[20] G. Ben-Dor, O. Igra, and T. Elperin (editors), *Handbook of Shock Waveas*, vol. 1 of *Volume 1: Theoretical, Experimental, and Numerical Techniques*, Academic Press, San Diego, San Francisco, New York (2001).

[21] D. J. Benson, Computational methods in Lagrangian and Eulerian hydrocodes, *Comput. Method Appl. M.*, 99 (1992), 235–394.

[22] K. Binder, *Monte Carlo and Molecular Dynamics simulations in polymer science*, Oxford University Press (1995).

[23] K. Binder, Applications of Monte Carlo methods to statistical physics, *Rep. Prog. Phys.*, 60 (1997), 487–559.

[24] K. Binder and D. W. Heermann, *Monte Carlo simulations in Statistical Physics*, Springer Verlag, Berlin, Heidelberg, New York, Tokio (1988).

[25] J. F. W. Bishop, C. R. Hill, and R. Hill, A theoretical derivation of the plastic properties of a polycrystalline face-centered material, *Philos. Mag.*, 42 (1951), 414–427.

[26] G. K. Bourov and A. Bhattacharya, Brownian dynamics simulation study of self-assembly of amphiphiles with large hydrophilic heads, *J. Chem. Phys.*, 122 (2005), 44702–1–44702–6.

[27] G. Brannigan, L. C. L. Lin, and F. L. H. Brown, Implicit solvent simulation models for biomembranes, *Eur. Biophys. J.*, 35 (2006), 104–124.

[28] G. Brannigan, P. F. Philips, and F. L. H. Brown, Flexible lipid bilayers in implicit solvent, *Phys. Rev. E*, 72 (2005), 011915–1–011915–4.

[29] W. Brendel, Shock Waves: A New Physical Principle in Medicine, *Eur. Surg. Res.*, 18 (1986), 177–180.

[30] H. S. Bridge, A. J. Lazarus, C. W. Snyder, E. J. Smith, L. Davis, P. J. Coleman, and D. E. Jones, Mariner V: Plasma and Magnetic Fields Observed near Venus, *Science*, 158 (1967), 1669–1673.

[31] E. M. Bringa, A. Caro, Y. M. Wang, M. Victoria, J. M. McNaney, B. A. Remington, R. F. Smith, B. R. Torralva, and H. Van Swygenhoven, Ultrahigh strength in nanocrystalline materials under shock loading, *Science*, 309 (2005), 1838–1841.

[32] E.-A. Brujan and Y. Matsumoto, Shock wave emission from a hemispherical cloud of bubbles in non-Newtonian fluids, *J. Non-Newton. Fluid*, 204 (2014), 32–37.

[33] M. J. Buehler, Multiscale aspects of mechanical properties of biological materials, *J. Mech. Behav. Biomed.*, 4 (2010), 125–127.

[34] M. Bulacu and E. van der Giessen, Effect of bending and torsion rigidity on self-diffusion in polymer melts: A molecular-dynamics study, *Journal of Chemical Physics*, 123 (2005), 114901-1–114901-13.

[35] J. C. Bushnell and D. J. McCloskey, Thermoelastic stress production in solids, *J. Appl. Phys.*, 39 (1968), 5541–5546.

[36] N. Cabibbo, Y. Iwasaki, and K. Schilling, High performance computing in lattice QCD, *Parallel Comput.*, 25 (1999), 1197–1198.

[37] W. Cai, A. Arsenlis, C. R. Weinberger, and V. V. Bulatov, A non-singular continuum theory of dislocations, *J. Mech. Phys. Solids*, 54 (2006), 561–587.

[38] R. Car and M. Parrinello, Unified approach for molecular dynamics and density-functional theory, *Phys. Rev. Lett.*, 55 (1985), 2471–2474.

[39] J. J. Chang, P. Engels, and M. A. Hoefer, Formation of dispersive shock waves by merging and splitting Bose-Einstein condensates., *Phys. Rev. Lett.*, 101 (2008), 170404-1–170404-4.

[40] R. Chang, G. S. Ayton, and G. A. Voth, Multiscale coupling of mesoscopic- and atomistic-level lipid bilayer simulations., *J. Chem. Phys.*, 122 (2005), 244716-1–244716-12.

[41] M. Chen, J. W. McCauley, and K. J. Hemker, Shock-Induced Localized Amorphization in Boron Carbide, *Science*, 299 (2003), 1563–1566.

[42] P. A. Cherenkov, Visible emission of clean liquids by action of gamma radiation, *R. Dokl. Akad. Nauk. SSSR*, 2 (1934), 451–454.

[43] S.-W. Chiu, H. L. Scott, and E. Jakobsson, A coarse-grained model based on Morse potential for water and n-alkanes, *J. Chem. Theory Comput.*, 6 (2010), 851–863.

[44] G. Ciccotti, G. Frenkel, and I. R. McDonald, *Simulation of Liquids and Solids*, vol. 16, North-Holland, Amsterdam (1987).

[45] A. Cohen, *Numerical Analysis*, McGraw-Hill, London (1962).

[46] I. R. Cooke and M. Deserno, Solvent-free model for self-assembling fluid bilayer membranes: Stabilization of the fluid phase based on broad attractive tail potentials, *J. Chem. Phys.*, 123 (2005), 224710-1–224710-13.

[47] R. Courant, Variational Methods for the Solution of Problems of Equilibrium and Vibrations, *Bull. Am. Math. Soc.*, 49 (1943), 1–23.

[48] C. C. Coussios, C. H. Farny, G. Ter Haar, and R. A. Roy, Role of acoustic cavitation in the delivery and monitoring of cancer treatment by high-intensity focused ultrasound (HIFU), *Int. J. Hyperthermia*, 23 (2007), 105–120.

[49] C. C. Coussios and R. A. Roy, Applications of Acoustics and Cavitation to Noninvasive Therapy and Drug Delivery, _Annual Review of Fluid Mechanics_, 40 (2008), 395–420.

[50] P. A. Cundall and O. D. Strack, A discrete numerical model for granular assemblies, _Geotechnique_, 29 (1979), 47–65.

[51] G. A. D'Addetta, F. Kun, and E. Ramm, On the application of a discrete model to the fracture process of cohesive granular materials, _Granul. Mat._, 4 (2002), 77–90.

[52] B. Damski, Shock waves in a one-dimensional Bose gas: From a Bose-Einstein condensate to a Tonks gas, _Phys. Rev. A_, 73 (2006), 043601-1–043601-10.

[53] M. S. Daw, Model of Metallic Cohesion: The Embedded-Atom Method, _Phys. Rev. B_, 39 (1988), 7441–7452.

[54] M. S. Daw and M. I. Baskes, Embedded-Atom Method - Derivation and Application to Impurities, Surfaces, and Other Defects in Metals, _Phys. Rev. B_, 29 (1984), 6443–6453.

[55] P. G. de Gennes, _Scaling concepts in polymer physics_, Cornell University Press, Ithaca, London (1979).

[56] B. Devincre, Three dimensional stress field expressions for straight dislocation segments, _Solid State Commun._, 93 (1995), 875–878.

[57] G. L. Dirichlet, Über die Reduction der positiven quadratischen Formen mit drei unbestimmten ganzen Zahlen, _J. Reine und Angew. Math._, 40 (1850), 209–227.

[58] M. Doi and S. F. Edwards, _The Theory of Polymer Dynamics_, Clarendon Press, Oxford (1986).

[59] A. G. Doukas and T. J. Flotte, Physical characteristics and biological effects of laser-induced stress waves, _Ultrasound Med. Biol._, 22 (1996), 151–164.

[60] A. G. Doukas and N. Kollias, Transdermal drug delivery with a pressure wave, *Adv. Drug Deliver Rev.*, 56 (2004), 559–579.

[61] A. G. Doukas, D. J. McAuliffe, S. Lee, V. Venugopalan, and T. J. Flotte, Physical factors involved in stress-wave-induced cell injury: The effect of stress gradient, *Ultrasound Med. Biol.*, 21 (1995), 961–967.

[62] T. Douki, S. Lee, K. Dorey, K. Dorey, T. J. Flotte, T. F. Deutsch, and A. G. Doukas, Stress-wave-induced injury to retinal pigment epithelium cells in vitro, *Lasers Surg Med*, 19 (1996), 249–259.

[63] J.-M. Drouffe, A. C. Maggs, S. Leibler, and S. Leibler, Computer simulations of self-assembled membranes, *Science*, 254 (1991), 1353–1356.

[64] B. Dünweg, D. Reith, M. Steinhauser, M. O. Steinhauser, and K. Kremer, Corrections to scaling in the hydrodynamic properties of dilute polymer solutions, *J. Chem. Phys.*, 117 (2002), 914–924.

[65] Z. Dutton, M. Budde, C. Slowe, and L. V. Hau, Observation of quantum shock waves created with ultra- compressed slow light pulses in a Bose-Einstein condensate., *Science*, 293 (2001), 663–668.

[66] E. W. Edwards, D. Wang, and H. Möhwald, Hierarchical Organization of Colloidal Particles: From Colloidal Crystallization to Supraparticle Chemistry, *Macromol. Chem. Phys.*, 208 (2007), 439–445.

[67] E. Egberts and H. J. C. Berendsen, Molecular dynamics simulation of a smectic liquid crystal with atomic detail, *J. Chem. Phys.*, 89 (1988), 3718.

[68] M. Elstner, D. Porezag, G. Jungnickel, J. Elsner, M. Haugk, T. Frauenheim, S. Suhai, and G. Seifert, Self-consistent-charge

density-functional tight-binding method for simulations of complex materials properties, *Phys. Rev. B*, 58 (1998), 7260–7268.

[69] H. Eslami and F. Müller-Plathe, How thick is the interphase in an ultrathin polymer film? Coarse-grained molecular dynamics simulations of polyamide-6,6 on graphene, *J. Phys. Chem.*, 117 (2013), 5249–5257.

[70] P. Español, Dissipative Particle Dynamics with energy conservation, *Europhys. Lett.*, 40 (1997), 631–636.

[71] P. Español and P. Warren, Statistical mechanics of dissipative particle dynamics, *Europhys. Lett.*, 30 (1995), 191–196.

[72] H. D. Espinosa and S. Lee, Modeling of ceramic microstructures: Dynamic damage initiation and evolution, in M. D. Furnish, L. C. Chhabildas, and R. S. Hixson (editors), *CP505, Shock Compression of Condensed Matter* (1999).

[73] H. G. Evertz, The loop algorithm, *Advances in Physics*, 52 (2003), 1–66.

[74] K. Falk, S. P. Regan, J. Vorberger, B. J. B. Crowley, S. H. Glenzer, S. X. Hu, C. D. Murphy, P. B. Radha, A. P. Jephcoat, J. S. Wark, D. O. Gericke, and G. Gregori, Comparison between x-ray scattering and velocity-interferometry measurements from shocked liquid deuterium, *Phys. Rev. E*, 87 (2013), 043112-1–043112-8.

[75] O. Farago, "Water-free" computer model for fluid bilayer membranes, *J. Chem. Phys.*, 119 (2003), 596–605.

[76] J. Fineberg, Materials science: close-up on cracks, *Nature*, 426 (2003), 131–132.

[77] M. W. Finnis and J. E. Sinclair, A simple empirical N-body potential for transition metals, *Philos. Mag. A*, 50 (1984), 45–55.

[78] V. Fock, Näherungsmethoden zur Lösung des quantenmechanischen Mehrkörperproblems, *Z. Phys.*, 61 (1932), 126–148.

[79] S. M. Foiles, M. I. Baskes, and M. S. Daw, Embedded-Atom-Method Functions for the Fcc Metals Cu, Ag, Au, Ni, Pd, Pt, and Their Alloys, *Phys. Rev. B*, 33 (1986), 7983–7991.

[80] J. W. Forbes, *Shock Wave Compression of Condensed Matter: A Primer*, Springer, New York, NY (2012).

[81] L. R. Forrest and M. S. Sansom, Membrane simulations: bigger and better?, *Curr. Opin. Struct. Biol.*, 10 (2000), 174–181.

[82] G. C. Ganzenmüller, S. Hiermaier, and M. O. Steinhauser, Shockwave induced damage in lipid bilayers: a dissipative particle dynamics simulation study, *Soft Matter*, 7 (2011), 4307–4317.

[83] G. C. Ganzenmüller, S. Hiermaier, and M. O. Steinhauser, Energy-based coupling of smooth particle hydrodynamics and molecular dynamics with thermal fluctuations, *Eur. Phys. J. Special Topics*, 206 (2012), 51–60.

[84] T. S. Gates, G. M. Odegard, S. J. V. Frankland, and T. C. Clancy, Computational materials: Multi-scale modeling and simulation of nanostructured materials, *Compos. Sci. Technol.*, 65 (2005), 2416–2434.

[85] T. C. Germann and K. Kadau, Trillion-atom molecular dynamics becomes a reality, *Int. J. Mod. Phys. C*, 19 (2008), 1315–1319.

[86] S. Ghosh and Y. Liu, Voronoi cell finite element model based on micropolar theory of thermoelasticity for heterogeneous materials, *Int. J. Numer. Meth. Engng.*, 38 (1995), 1361–1398.

[87] R. Goetz, G. Gompper, and R. Lipowsky, Mobility and elasticity of self-assembled membranes, *Phys. Rev. Lett.*, 82 (1999), 221–224.

[88] R. Goetz and R. Lipowsky, Computer simulations of bilayer membranes: Self-assembly and interfacial tension, *J. Chem. Phys.*, 108 (1998), 7397–7409.

[89] K. Grass, A. Blumen, M. O. Steinhauser, and K. Thoma, Sequential modeling of failure behavior in cohesive brittle materials, in R. García-Rocho, H. J. Herrmann, and M. Sean (editors), *The 5th Int. Conference on Micromechanics of Granular Media, Stuttgart, Germany, 18.-22. July 2005*, pages 1447–1550 (2005).

[90] T. Gröger, U. Tüzün, and D. M. Heyes, Modelling and measuring of cohesion in wet granular materials, *Powder Technol.*, 133 (2003), 203–215.

[91] M. Guenza, Cooperative dynamics in semiflexibile unentangled polymer fluids, *Journal of Chemical Physics*, 119 (2003), 7568–7578.

[92] D. A. Gurnett and W. S. Kurth, Electron plasma oscillations upstream of the solar wind termination shock, *Science*, 309 (2005), 2025–2027.

[93] D. A. Gurnett and W. S. Kurth, Intense plasma waves at and near the solar wind termination shock, *Nature*, 454 (2008), 78–80.

[94] E. Guyon, J.-P. Hulin, L. Petit, and C. D. Mitescu, *Physical Hydrodynamics*, Oxford University Press (2001).

[95] J. P. Hansen and I. R. McDonald, *Theory of Simple Liquids*, Academic Press (2005).

[96] L. Harnau, R. G. Winkler, and P. Reineker, Influence of stiffness on the dynamics of macromolecules in a melt, *J. Chem. Phys.*, 106 (1997), 2469–2476.

[97] L. Harnau, R. G. WInkler, and P. Reineker, On the dynamics of polymer melts: Contribution of Rouse and bending modes, *Europhys. Lett.*, 45 (1999), 488–494.

[98] G. R. Harris, R. C. Preston, and A. S. Dereggi, The impact of piezoelectric PVDF on medical ultrasound exposure measurements, standards, and regulations, *IEEE Trans. Ultrason., Ferroelect., Freq. Contr.*, 47 (1999), 1321–1335.

[99] R. A. Harris and J. E. Hearst, On Polymer Dynamics, *J. Chem. Phys.*, 44 (1966), 2595–2602.

[100] D. R. Hartree, The Wave Mechanics of an Atom with a non-Coulomb Central Field. Part III. Term Values and Intensities in Series in Optical Spectra, *Math. Proc. Cambridge*, 24 (1928), 426–437.

[101] J. E. Hearst and R. A. Harris, On Polymer Dynamics. III. Elastic Light Scattering, *J. Chem. Phys.*, 46 (1967), 398–398.

[102] P. Hohenberg and W. Kohn, Inhomogeneous electron gas, *Phys. Rev.*, 36 (1964), 864–871.

[103] B. L. Holian, Molecular dynamics comes of age for shockwave research, *Shock Waves*, 13 (2004), 489–495.

[104] B. L. Holian and P. S. Lomdahl, Plasticity induced by shock waves in nonequilibrium molecular-dynamics simulations, *Science*, 280 (1998), 2085–2088.

[105] E. A. Holm and C. C. Battaile, The computer simulation of microstructural evolution, *JOM*, 53 (2001), 20–23.

[106] P. J. Hoogerbrugge and J. M. V. A. Koelman, Simulating microscopic hydrodynamic phenomena with dissipative particle dynamics, *Europhys. Lett.*, 19 (1992), 155–160.

[107] W. G. Hoover, *Smooth Particle Applied Mechanics: The State of the Art*, World Scientific Publishing Co., Inc, Singapore, advanced series in nonlinear dynamics edn. (2006).

[108] M.-J. Huang, R. Kapral, A. S. Mikhailov, and H.-Y. Chen, Coarse-grain model for lipid bilayer self-assembly and dynamics: Multiparticle collision description of the solvent, *J. Chem. Phys.*, 137 (2012), 055101-1–055101-10.

[109] W.-X. Huang, C. B. Chang, and H. J. Sung, Three-dimensional simulation of elastic capsules in shear flow by the penalty immersed boundary method, *J. Comput. Phys.*, 231 (2012), 3340–3364.

[110] H. Hugoniot, *Mémoire sur la propagation du mouvement dans un fluide indéfini (seconde Partie)*, vol. 4, Journal de Mathématiques Pures et Appliquées (1888).

[111] S. Iakovlev, S. Iakovlev, C. Buchner, C. Buchner, B. Thompson, A. Lefieux, and A. Lefieux, Resonance-like phenomena in a submerged cylindrical shell subjected to two consecutive shock waves: The effect of the inner fluid, *J. Fluids Struct.*, 50 (2014), 153–170.

[112] L. D. Johns, Nonthermal effects of therapeutic ultrasound: the frequency resonance hypothesis., *J Athl Train*, 37 (2002), 293–299.

[113] T. Juhasz, X. H. Hu, L. Turi, and Z. Bor, Dynamics of shock waves and cavitation bubbles generated by picosecond laser pulses in corneal tissue and water, *Lasers Surg. Med.*, 15 (1993), 91–98.

[114] T. Juhasz, G. A. Kastis, C. Suárez, Z. Bor, and W. E. Bron, Time-resolved observations of shock waves and cavitation bubbles generated by femtosecond laser pulses in corneal tissue and water, *Lasers Surg. Med.*, 19 (1995), 23–31.

[115] D. Kadau, G. Bartels, L. Brendel, and D. E. Wolf, Contact dynamics simulations of compacting cohesive granular systems, *Comp. Phys. Comm.*, 147 (2002), 190–193.

[116] K. Kadau, T. C. Germann, P. S. Lomdahl, R. C. Albers, J. S. Wark, A. Higginbotham, and B. L. Holian, Shock waves in polycrystalline iron, *Phys. Rev. Lett.*, 98 (2007), 135701-1–135701-4.

[117] K. Kadau, T. C. Germann, P. S. Lomdahl, and B. L. Holian, Microscopic view of structural phase transitions induced by shock waves, *Science*, 296 (2002), 1681–1684.

[118] U. D. Kahlert, G. Nikkhah, and J. Maciaczyk, Epithelial-to-mesenchymal(-like) transition as a relevant molecular event in malignant gliomas, *Cancer Lett.*, 331 (2013), 131–138.

[119] H. A. Karimi-Varzaneh and F. Müller-Plathe, Coarse-Grained Modeling for Macromolecular Chemistry , in B. Kirchner and J. Vrabec (editors), *Topics in Current Chemistry*, pages 326–321, Springer, Berlin, Heidelberg (2011).

[120] M. D. Knudson, M. P. Desjarlais, and D. H. Dolan, Shock-Wave Exploration of the High-Pressure Phases of Carbon, *Science*, 322 (2008), 1822–1825.

[121] T. Kodama, A. G. Doukas, and M. R. Hamblin, Shock wave-mediated molecular delivery into cells, *Biochim. Biophys. Acta*, 1542 (2002), 186–194.

[122] T. Kodama, T. Kodama, M. R. Hamblin, M. R. Hamblin, and A. G. Doukas, Cytoplasmic molecular delivery with shock waves: importance of impulse, *Biophys. J.*, 79 (2000), 1821–1832.

[123] W. Kohn, Density functional and density matrix method scaling linearly with the number of atoms, *Phys. Rev. Lett.*, 76 (1996), 3168–3171.

[124] W. Kohn and L. J. Sham, Self-Consistent Equations Including Exchange and Correlation Effects, *Phys. Rev.*, 140 (1965), A1133–A1138.

[125] K. Koshiyama, T. Kodama, T. Yano, and S. S. Fujikawa, Structural Change in Lipid Bilayers and Water Penetration Induced by Shock Waves: Molecular Dynamics Simulations, *Biophys. J.*, 91 (2006), 2198–2205.

[126] K. Koshiyama, T. Kodama, T. Yano, and S. S. Fujikawa, Molecular dynamics simulation of structural changes of lipid bilayers induced by shock waves: Effects of incident angles, *Biochim. Biophys. Acta*, 1778 (2008), 1423–1428.

[127] O. Kratky and G. Porod, Röntgenuntersuchung gelöster Fadenmoleküle, *Rec. Trv. Chim*, 68 (1949), 1106–1115.

[128] T. Kreer, J. Baschnagel, M. Mueller, and K. Binder, Monte Carlo Simulation of long chain polymer melts: Crossover from Rouse to reptation dynamics, *Macromol.*, 34 (2001), 1105–1117.

[129] P. O. K. Krehl, *History of Shock Waves, Explosions and Impact: A Chronological and Biographical Reference*, Springer, Berlin (2009).

[130] A. Krell, P. Blank, H. Ma, T. Hutzler, M. P. B. van Bruggen, and R. Apetz, Transparentsintered corundum with high hardness and strength, *J. Am. Ceram. Soc.*, 1 (2003), 12–18.

[131] K. Kremer and F. Müller-Plathe, Multiscale simulation in polymer science, *Mol. Simul.*, 28 (2010), 729–750.

[132] S. Krushev, W. Paul, and G. D. Smith, The role of internal rotational barriers in polymer melt chain dynamics, *Macromol.*, 35 (2002), 4198–4203.

[133] M. Kühn and M. O. Steinhauser, Modeling and simulation of microstructures using power diagrams: Proof of the concept, *Appl. Phys. Lett.*, 93 (2008), 034102–1–034102–3.

[134] A. Ladd, Short-time motion of colloidal particles: numerical simulation via a fluctuating lattice-Boltzmann equation, *Phys. Rev. Lett.*, 70 (1993), 1339–1342.

[135] L. D. Landau and E. M. Lifshitz, *Statistical Physics*, vol. 5 of *Course of Theoretical Physics*, Elsevier, Oxford, MA, 3 edn. (1980).

[136] L. D. Landau and E. M. Lifshitz, *Fluid Mechanics*, Course of theoretical physics, Pergamon Press, Oxford, New York, Bejing, Frankfurt,Sydney,Tokyo, Toronto, 2 edn. (1987).

[137] E. M. Lauridsen, Approaches for 3D Materials characterization, *JOM*, 58 (2006), 12.

[138] J. Lechuga, D. Drikakis, and S. Pal, Molecular dynamics study of the interaction of a shock wave with a biological membrane, *Int. J. Numer. Mech. Fluids*, 57 (2008), 677–692.

[139] C.-H. Lee, W.-C. Lin, and J. Wang, All-optical measurements of the bending rigidity of lipid-vesicle membranes across structural phase transitions, *Phys. Rev. E*, 64 (2001), 020901–1–02091–4.

[140] S. Lee, T. Anderson, H. Zhang, T. J. Flotte, and A. G. Doukas, Alteration of cell membrane by stress waves in vitro, *Ultrasound Med. Biol.*, 22 (1996), 1285–1293.

[141] S. Lee and A. G. Doukas, Laser-generated stress waves and their effects on the cell membrane, *IEEE J. Sel. Top. Quant.*, 5 (1999), 997–1003.

[142] J. S. Leszczynski, A discrete model of a two-particle contact applied to cohesive granular materials, *Granul. Mat.*, 5 (2003), 91–98.

[143] A. C. Lewis, C. Suh, M. Stukowski, and A. B. Geltmacher, Tracking correlations between mechanical response and microstructure in three-dimensional reconstructions of a commercial stainless steel, *Scripta Materialia*, 58 (2008), 575–578.

[144] A. C. Lewis, S. Suh, M. Stukowski, A. B. Geltmacher, G. Spanos, and K. Rajan, Quantitative analysis and feature recognition in 3-D microstructural data sets, *JOM*, 12 (2006), 51–56.

[145] H. W. Liepmann and A. Roshko, *Elements of Gasdynamics*, John Wiley & Sons, New York (1957).

[146] M. Lindau and G. Alvarez de Toledo, The fusion pore, *Biochim. Biophys. Acta*, 1641 (2003), 167–173.

[147] J. E. Lingeman, J. A. McAteer, E. Gnessin, and A. P. Evan, Shock wave lithotripsy: advances in technology and technique, *Nat. Rev. Urol.*, 6 (2009), 660–670.

[148] R. Lipowsky, Biomimetic membrane modelling: pictures from the twilight zone, *Nat. Mater.*, 3 (2004), 589–591.

[149] W. K. Liu, S. Hao, T. Belytschko, and S. F. Li, Multiple scale meshfree methods for damage fracture and localization, *Comput. Mater. Sci.*, 16 (1999), 197–205.

[150] L. B. Lucy, A numerical approach to the testing of the fission hypothesis, *J. Astronom.*, 82 (1977), 1013–1024.

[151] J. Lützen, *The Prehistory of the Theory of Distributions*, Springer, New York, Heidelberg, Berlin (1982).

[152] A. P. Lyubartsev, Multiscale modeling of lipids and lipid bilayers, *Eur. Biophys. J.*, 35 (2005), 53–61.

[153] E. Mach and P. Salcher, Photographische Fixirung der durch Projectile in der Luft eingeleiteten Vorgänge, *Ann. Phys.*, 268 (1887), 277–291.

[154] D. Majumder and A. Mukherjee, Multi-scale modeling approaches in systems biology towards the assessment of cancer treatment dynamics: adoption of middle-out rationalist approach, *Adv. Cancer Res.*, 2013 (2013), 1–26.

[155] S. J. Marrink, H. J. Risselada, S. Yefimov, D. P. Tieleman, and A. H. de Vries, The MARTINI force field: Coarse grained model for biomolecular simulations, *J. Phys. Chem.*, 111 (2007), 7812–7824.

[156] A. Mayer, Membrane Fusion in Eukaryotic Cells, *Annu. Rev. Cell. Dev. Biol.*, 18 (2002), 289–314.

[157] S. McClure and C. Dorfmüller, Extracorporeal shock wave therapy: Theory and equipment, *Clin. Tech. Equine Pract.*, 2 (2002), 348–357.

[158] C. F. McKee and B. T. Draine, Interstellar shock waves, *Science*, 252 (1991), 397–403.

[159] P. L. McNeil and M. Terasaki, Coping with the inevitable: how cells repair a torn surface membrane, *Nat. Cell Biol.*, 3 (2001), E124–E129.

[160] N. Metropolis, The beginning of the Monte Carlo method, *Los Alamos Science*, (1987).

[161] N. Metropolis, A. W. Rosenbluth, M. N. Rosenbluth, A. H. Teller, and E. Teller, Equation of State Calculations by Fast Computing Machines, *J. Chem. Phys.*, 21 (1953), 1087–1092.

[162] N. Metropolis and S. Ulam, The Monte Carlo method, *J. am. Stat. Assoc.*, 44 (1949), 335–341.

[163] M. Millot, N. Dubrovinskaia, A. Černok, S. Blaha, L. S. Dubrovinsky, D. G. Braun, P. M. Celliers, G. W. Collins, J. H. Eggert, and R. Jeanloz, Planetary science. Shock compression of stishovite and melting of silica at planetary interior conditions., *Science*, 347 (2015), 418–420.

[164] S. S. Mokrushin, N. B. Anikin, and S. N. Malyugina, An interferometer with time-and-frequency signal compression for studying properties of materials in shock wave experiments, *Instruments and ...*, (2014), 083108-1–083108-6.

[165] J. J. Monaghan, An introduction to SPH, *Comp. Phys. Comm.*, 48 (1988), 89–96.

[166] J. J. Monaghan, Smoothed particle hydrodynamics, *Ann. Rev. Astron. Astrophys.*, 30 (1992), 543–574.

[167] S. E. Mulholland, S. Lee, D. J. McAuliffe, and A. G. Doukas, Cell loading with laser-generated stress waves: The role of the stress gradient, *Pharm. Res.*, 16 (1999), 514–518.

[168] F. Müller-Plathe, Coarse-graining in polymer simulation: from the atomistic to the mesoscopic scale and back, *Chem. Phys. Chem.*, 3 (2002), 755–769.

[169] P. Murray, C. Edwards, M. Tindall, and P. Maini, From a discrete to a continuum model of cell dynamics in one dimension, *Phys. Rev. E*, 80 (2009), 031912-1–031912-10.

[170] S. O. Nielsen, C. F. Lopez, G. Srinivas, and M. L. Klein, Coarse grain models and the computer simulation of soft materials, *J. Phys.-Condens. Mat.*, 16 (2004), 481–R512.

[171] H. Noguchi, Membrane simulation models from nanometer to micrometer scale, *J. Phys. Soc. Jpn.*, 78 (2009), 041007-1–041007-9.

[172] H. Noguchi, Solvent-free coarse-grained lipid model for large-scale simulations, *J. Chem. Phys.*, 134 (2011), 055101–055101.

[173] H. Noguchi and M. Takasu, Self-assembly of amphiphiles into vesicles: A Brownian dynamics simulation, *Phys. Rev. E*, 64 (2001), 041913-1–041913-7.

[174] A. Okabe, B. Boots, and K. Sugihara, *Spatial Tesselations - Concepts and Applications of Voronoi Diagrams*, John Wiley and Sons (1992).

[175] K. Olbrich, W. Rawicz, D. Needham, and E. Evans, Water permeability and mechanical strength of polyunsaturated lipid bilayers., *Biophys. J.*, 79 (2000), 321–327.

[176] M. Orsi, J. Michel, and J. W. Essex, Coarse-grain modelling of DMPC and DOPC lipid bilayers., *J. Phys.-Condens. Mat.*, 22 (2010), 155106-1–155106–15.

[177] S. A. Pandit and H. L. Scott, Multiscale simulations of heterogeneous model membranes., *Biochim. Biophys. Acta*, 1788 (2009), 136–148.

[178] M. Pasquali, V. Shankar, and D. C. Morse, Viscoelasticity of dilute solutions of semiflexible polymers, *Phys. Rev. E*, 64 (2001), 020802–1–020802–4.

[179] W. Paul, W. Paul, D. Y. Yoon, D. O. Yoon, G. D. Smith, and G. D. Smith, An Optimized United Atom Model for Simulations of Polymethylene Melts, *J. Chem. Phys.*, 103 (1995), 1702–1709.

[180] W. Paul, G. D. Smith, and D. Y. Yoon, Static and dynamic properties of an-C100H202 melt from molecular dynamics simulations, *Macromol.*, 30 (1997), 7772–7780.

[181] F. G. Pazzona and P. Demontis, A grand-canonical Monte Carlo study of the adsorption properties of argon confined in ZIF-8: local thermodynamic modeling, *J. Phys. Chem.*, 117 (2012), 349–357.

[182] R. R. Phillips, *Crystals, defects and microstructures*, Modeling across scales, Cambridge Univ. Press, Cambridge (2002).

[183] T. Piran (editor), *Statistical Mechanics of Membranes and Interfaces*, World Scientific Publishing Co., Inc, 2 edn. (2004).

[184] S. Pogodin and V. A. Baulin, Coarse-grained models of phospholipid membranes within the single chain mean field theory, *Soft Matter*, 6 (2010), 2216–2226.

[185] R. B. Potts and C. Domb, Some generalized order-disorder transformations, *Math. Proc. Cambridge*, 48 (1952), 106–109.

[186] M. Praprotnik, L. D. Site, and K. Kremer, Multiscale simulation of soft matter: From scale bridging to adaptive resolution, *Annu. Rev. Phys. Chem.*, 59 (2008), 545–571.

[187] Y. P. Raizer and Y. B. Zel'dovich, *Physics of shock waves and high-temperature hydrodynamic phenomena*, Academic Press, New York, NY (1967).

[188] W. M. Rankine, On the Thermodynamic Theory of Waves of Finite Longitudinal Disturbances, *Phil. Trans. R. Soc. Lond.*, 160 II (1870), 277–288.

[189] D. C. Rapaport, *The Art of Molecular Dynamics Simulation*, Cambridge University Press, Cambridge (2004).

[190] F. H. Ree, Analytic representation of thermodynamic data for the Lennard-Jones fluid, *J. Chem. Phys.*, 73 (1980), 5401–5403.

[191] B. Riemann, Über die Fortpflanzung ebener Luftwellen von endlicher Schwingungsweite, *Tech. rep.*, Ges. d. Wiss (1860).

[192] V. Rodriguez, R. Saurel, G. Jourdan, and L. Houas, Solid-particle jet formation under shock-wave acceleration, *Phys. Rev. E*, 88 (2013), 063011.

[193] D. Saari, A visit to the Newtonian N-body problem via elementary complex variables, *Am. Math. Monthly*, 89 (1990), 105–119.

[194] M. H. Sadd, Q. M. Tai, and A. Shukla, Contact law effects on wave propagation in particulate materials using distinct element modeling, *Int. J. Non-Linear Mechanics*, 28 (1993), 251–265.

[195] A. Saidi, S. Javerzat, A. Bellahcène, J. De Vos, L. Bello, V. Castronovo, M. Deprez, H. Loiseau, A. Bikfalvi, and M. Hagedorn, Experimental anti-angiogenesis causes upregulation of genes associated with poor survival in glioblastoma, *Int. J. Cancer*, 122 (2008), 2187–2198.

[196] T. Schindler, D. Kröner, and M. O. Steinhauser, On the dynamics of molecular self-assembly and the structural analysis of bilayer membranes using coarse-grained molecular dynamics simulations, *Biochim. Biophys. Acta*, 1858 (2016), 1955–1963.

[197] M. Schmidt, U. Kahlert, J. Wessolleck, D. Maciaczyk, B. Merkt, J. Maciaczyk, J. Osterholz, G. Nikkhah, and M. O. Steinhauser, Characterization of a setup to test the impact of high-amplitude pressure waves on living cells, *Nature Sci. Rep.*, 4 (2014), 3849-1–3849–9.

[198] L. Schwartz, *Théorie des Distributions*, vol. I, II, Herrmann et Cie, Paris (1950).

[199] U. Seifert, Configurations of fluid membranes and vesicles, *Advances in Physics*, 46 (1997), 13–137.

[200] S. J. Singer and G. L. Nicolson, The fluid mosaic model of the structure of cell membranes., *Science*, 175 (1972), 720–731.

[201] V. Singh, A. Madan, H. Suneja, and D. Chand, Propagation of spherical shock waves in water, *Proc. Indian Acad. Sci.*, 3 (1980), 169–175.

[202] S. W. I. Siu, R. Vácha, P. Jungwirth, and R. A. Böckmann, Biomolecular simulations of membranes: physical properties from different force fields, *J. Phys. Chem.*, 128 (2008), 125103-1–125103–12.

[203] S. L. Sobolev, Méthode nouvelle à resoudre le problème de Cauchy pour les équations linéaires hyperboliques normales, *Mat. Sbornik*, 1 (1936), 39–72.

[204] G. A. Sod, A survey of several finite difference methods for systems of nonlinear hyperbolic conservation laws, *J. Comput. Phys.*, 27 (1978), 1–31.

[205] H. Sprong, P. van der Sluijs, and G. van Meer, How proteins move lipids and lipids move proteins, *Nat. Rev. Mol. Cell Biol.*, 2 (2001), 504–513.

[206] M. O. Steinhauser, A molecular dynamics study on universal properties of polymer chains in different solvent qualities. Part I. A review of linear chain properties, *J. Chem. Phys.*, 122 (2005), 094901-1–094901-13.

[207] M. O. Steinhauser, Computational methods in polymer physics, *Recent Res. Devel. Phys.*, 7 (2006), 59–97.

[208] M. O. Steinhauser, Static and dynamic scaling of semiflexible polymer chains—a molecular dynamics simulation study of single chains and melts, *Mech. Time-Depend. Mater.*, 12 (2008), 291–312.

[209] M. O. Steinhauser, *Introduction to Molecular Dynamics Simulations: Applications in Hard and Soft Condensed Matter Physics*, In: Molecular Dynamics - Studies of Synthetic and Biological Macromolecules, Prof. Lichang Wang (ed.), DOI: 10.5772/36289. InTech (2012).

[210] M. O. Steinhauser, *Computer Simulation in Physics and Engineering*, deGruyter, Berlin, Boston, 1st edn. (2013).

[211] M. O. Steinhauser, *Computational Multiscale Modeling of Fluids and Solids - Theory and Applications*, Springer, Berlin, Heidelberg, New York, 2 edn. (2016).

[212] M. O. Steinhauser, On the Destruction of Cancer Cells Using Laser-Induced Shock-Waves: A Review on Experiments and Multiscale Computer Simulations, *Radiol. Open J.*, 1 (2016), 60–75.

[213] M. O. Steinhauser, Multiscale modeling, coarse-graining and shock wave computer simulations in materials science, *AIMS Materials Science*, 4 (2017), 1319–1357.

[214] M. O. Steinhauser, *Quantenmechanik für Naturwissenschaftler,* Ein Lehr- und Übungsbuch mit zahlreichen Aufgaben und Lösungen, Springer, Berlin, Heidelberg, 1st edn. (2017).

[215] M. O. Steinhauser, K. Grass, E. Strassburger, and A. Blumen, Impact failure of granular materials – Non-equilibrium multiscale simulations and high-speed experiments, *Int. J. Plast.*, 25 (2009), 161–182.

[216] M. O. Steinhauser, K. Grass, K. Thoma, and A. Blumen, Impact dynamics and failure of brittle solid states by means of nonequilibrium molecular dynamics simulations, *Europhys. Lett.*, 73 (2006), 62–68.

[217] M. O. Steinhauser and S. Hiermaier, A Review of Computational Methods in Materials Science: Examples from Shock-Wave and Polymer Physics, *Int. Mol. Sci.*, 10 (2009), 5135–5216.

[218] M. O. Steinhauser and M. Kühn, Modeling of Shock-Wave Failure in Brittle Materials, in P. Gumbsch (editor), *MMM: The 3rd International Conference on Muliscale Materials Modeling, Freiburg, Germany, 18.-22. 09. 2005*, pages 1–3 (2006).

[219] M. O. Steinhauser and M. Kühn, The use of optimized power-diagrams for mesoscopic shock wave modeling, in *The 13th International Symposium of Plasticity And Its Current Applications, Anchorage, Alaska, 2-6 June 2007*, pages 1–3 (2007).

[220] M. O. Steinhauser, M. Kühn, and K. Grass, Numerical simulation of fracture and failure dynamics in brittle solids, in *The 12th Int Symposium on Plasticity and Its Current Applications, Halifax, Nova Scotia, Canada, 17-22 July 2006*, pages 1–3 (2006).

[221] M. O. Steinhauser and T. Schindler, Particle-based simulations of bilayer membranes: self-assembly, structural analysis, and shock-wave damage, *Comp. Part. Mech.*, (2016), 1–18.

[222] M. O. Steinhauser, J. Schneider, and A. Blumen, Simulating dynamic crossover behavior of semiflexible linear polymers in solution and in the melt, *J. Chem. Phys.*, 130 (2009), 164902-1–164902-8.

[223] M. O. Steinhauser, M. O. Steinhauser, and M. Schmidt, Destruction of cancer cells by laser-induced shock waves: recent developments in experimental treatments and multiscale computer simulations, *Soft Matter*, 10 (2014), 4778–4788.

[224] M. J. Stevens, Coarse-grained simulations of lipid bilayers, *J. Chem. Phys.*, 121 (2004), 11942–11948.

[225] G. Stoltz, A reduced model for shock and detonation waves. I. The inert case, *Europhys. Lett.*, 76 (2006), 849–855.

[226] D. Stoyan, W. S. Kendall, and J. Mecke, *Stochastic Geometry and its Applications*, John Wiley & Sons, Inc., Chichester (1987).

[227] A. P. Sutton, M. W. Finnis, D. G. Pettifor, and Y. Ohta, The tight-binding bond model, *J. Phys. C: Solid State Phys.*, 21 (1988), 35–66.

[228] V. Tozzini, Coarse-grained models for proteins, *Curr. Opin. Struct. Biol.*, 15 (2004), 144–150.

[229] B. D. unweg, B. Duenweg, D. Reith, M. O. Steinhauser, D. Reith, and K. Kremer, Corrections to Scaling in the Hydrodynamic Properties of Dilute Polymer Solutions, *J. Chem. Phys.*, 117 (2002), 914.

[230] G. van Meer, D. R. Voelker, and G. W. Feigenson, Membrane lipids: where they are and how they behave, *Nat. Rev. Mol. Cell Biol.*, 9 (2008), 112–124.

[231] I. Vattulainen, M. Karttunen, G. Besold, and J. M. Polson, Integration schemes for Dissipative Particle Dynamics Simulations: From softly interacting systems towards hybrid models, *J. Chem. Phys.*, 116 (2002), 3967–3979.

[232] V. Venugopalan and T. F. Deutsch, Stress generated in polyimide by excimer-laser irradiation, *J. Appl. Phys.*, 74 (1993), 4181–4189.

[233] A. Vogel, S. Busch, and U. Parlitz, Shock wave emission and cavitation bubble generation by picosecond and nanosecond optlical breakdown in water, *J. Acoust. Soc. Am.*, 100 (1996), 148–165.

[234] J. M. Walsh and M. H. Rice, Dynamic compression of liquids from measurements on strong shock waves, *J. Chem. Phys.*, (1957), 815–823.

[235] O. R. Walton and R. L. Braun, Viscosity, granular-temperature, and stress calculations for shearing assemblies of inelastic, frictional disks, *J. Rheol.*, 30 (1986), 949–980.

[236] C.-J. Wang, An overview of shock wave therapy in musculoskeletal disorders, *Chang Gung Med J*, 26 (2003), 220–232.

[237] S. Wang, V. Frenkel, and V. Zderic, Optimization of pulsed focused ultrasound exposures for hyperthermia applications., *J. Acoust. Soc. Am.*, 130 (2011), 599–609.

[238] Y. Wang, J. K. Sigurdsson, E. Brandt, and P. J. Atzberger, Dynamic implicit-solvent coarse-grained models of lipid bilayer membranes: fluctuating hydrodynamics thermostat, *Phys. Rev. E*, 88 (2013), 023301-1–023301-5.

[239] Z. B. Wang, J. Wu, L. Q. Fang, H. Wang, F. Q. Li, Y. B. Tian, X. B. Gong, H. Zhang, L. Zhang, and R. Feng, Preliminary ex vivo feasibility study on targeted cell surgery by high intensity focused ultrasound (HIFU), *Ultrasonics*, 51 (2011), 369–375.

[240] Z.-J. Z. Wang and M. Deserno, A systematically coarse-grained solvent-free model for quantitative phospholipid bilayer simulations, *J. Phys. Chem. B*, 114 (2010), 11207–11220.

[241] A. Warshel and M. Levitt, Theoretical studies of enzymic reactions: Dielectric, electrostatic and steric stabilization of the carbonium ion in the reaction of lysozyme, *J. Mol. Biol.*, 103 (1976), 227–249.

[242] W. Weibull, A statistical distribution function of wide applicability, *J. Appl. Mech.*, 18 (1951), 293–297.

[243] S. J. Weiner, P. A. Kollman, D. A. Case, U. C. Singh, C. Ghio, G. Alagona, S. Profeta, and P. Weiner, A new force field for molecular mechanical simulation of nucleic acids and proteins, *J. Am. Chem. Soc.*, 106 (1984), 765–784.

[244] R. G. Winkler, M. O. Steinhauser, and P. Reineker, Complex formation in systems of oppositely charged polyelectrolytes: a molecular dynamics simulation study., *Phys. Rev. E*, 66 (2002), 021802-1–021802-7.

[245] S. Wolfram, Undecidability And Intractability in Theoretical Physics, *Phys. Rev. Lett.*, 54 (1985), 735–738.

[246] C. J. Woods and A. J. Mulholland, Multiscale modelling of biological systems, *Chem. Modell.*, 5 (2008), 13–50.

[247] Xiao and T. Belytschko, A bridging domain method for coupling continua with molecular dynamics, *Comput. Method Appl. M.*, 193 (2004), 25–25.

[248] X. P. Xu and A. Needleman, Numerical simulations of dynamic interfacial crack growth allowing for crack growth away from the bond line, *Int. J. Fracture*, 74 (1996), 253–275.

[249] M. Yamalidou, Molecular ideas in hydrodynamics, *Annals of Science*, 55 (1998), 369–400.

[250] S. Yang and J. Qu, Coarse-grained molecular dynamics simulations of the tensile behavior of a thermosetting polymer, *Phys. Rev. E*, 90 (2014), 012601-1–012601-8.

[251] W. Yang, J. Cai, Y. S. Ing, and C. C. Mach, Transient Dislocation Emission From a Crack Tip, *J. Mech. Phys. Solids*, 49 (2001), 2431–2453.

[252] S. Yip (editor), *Handbook of materials modeling*, Springer, Berlin (2005).

[253] H. Y. Yoshikawa, J. Cui, K. Kratz, T. Matsuzaki, S. Nakabayashi, A. Marx, U. Engel, A. Lendlein, and M. Tanaka, Quantitative evaluation of adhesion of osteosarcoma cells to hydrophobic polymer substrate with tunable elasticity, *J. Phys. Chem. B*, 116 (2012), 8024–8030.

[254] H. Yuan, C. Huang, J. Li, G. Lykotrafitis, and S. Zhang, One-particle-thick, solvent-free, coarse-grained model for biological and biomimetic fluid membranes, *Phys. Rev. E*, 82 (2010), 011905-1–011905-8.

[255] P. D. Zavattieri and H. D. Espinosa, An examination if the competition between bulk behavior and interfacial behavior of ceramics subjected to dynamics pressure-shear loading, *J. Mech. Phys. Solids*, 51 (2003), 607–635.

[256] X. Zhang, J. Guanghui, and H. Huang, Fragment identification and statistics method of hypervelocity impact SPH simulation, *Chin. J. Aeronaut.*, 24 (2011), 18–24.

[257] J. Zheng, Q. F. Chen, Y. J. Gu, and Z. Y. Chen, Hugoniot measurements of double-shocked precompressed dense xenon plasmas, *Phys. Rev. E*, 86 (2012), 066406-1–066406-5.

Index

© Springer Fachmedien Wiesbaden Gmbh 2018
M. O. Steinhauser, *Multiscale Modeling and Simulation of Shock Wave-Induced Failure in Materials Science*,
https://doi.org/10.1007/978-3-658-21134-9

Printed in the United States
By Bookmasters